Activities

The activities section presents some interesting and relevant scientific stories for you to look at and use your skills to answer the questions.

Questions test a range of essential skills including Working Scientifically, Maths Skills and Writing Extended Answers.
Easy cross references let you look up explanations of the skills needed to answer questions you are struggling with.

HIGHER The 'Higher' icon flags explanations and questions that will only be tested on the Higher Tier route through a GCSE course.

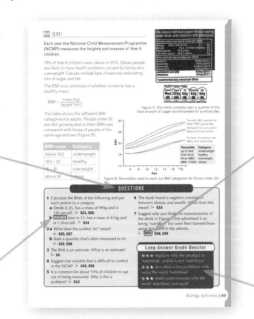

WEA Where the WEA (Writing Extended Answers) icon appears next to a question in this book, you will be expected to answer the question using full sentences, good organisation and with correct spelling and punctuation.

Dedicated questions at the end of many activities help you prepare for writing longer answers in exams – there is a booster box to give you hints on how to give a good answer.

Answers

There are answers to all the questions within the book so that you can easily check your progress. Many of the answers have tips on how you could improve your answers or warnings about mistakes that are often made. So, they're worth reading carefully.

Some answers, with this icon, show you what needs to be done to an answer to make it even better.

The worked examples showing good and poor responses to some long-answer questions will help you see how you can improve your own answers.

For the teacher

This book is designed to focus on the skills needed for GCSE science, rather than 'content knowledge'. It covers all the skills required by all the different awarding bodies for GCSE in the UK. We have also included some skills that are not formally required but which are useful in science and may be met by students when doing independent research.

The first section of the book presents the skills that need to be developed and the second section contains ten scientific contexts for each of biology, chemistry and physics. These contexts are not necessarily related to GCSE specification content. They mainly provide a wide range of interesting contexts in which students can apply their skills.

The book is designed to be self-contained and flexible, allowing its use to be dictated by the teacher rather than the book imposing a strict, formulaic approach on the way in which skills development is taught.

It is envisaged that most teachers will use the book for intervention. There are a number of ways in which this might be done.

- When students start work on their GCSE courses, it often soon becomes apparent that some students are lacking certain skills. This book allows the students to look specifically at those skills (on pages 11–52). Once covered, they can then practise applying these skills in a wide range of biological, chemical and physical contexts, helping to reinforce the learning and to identify any further areas of weakness/ misunderstanding.

- If you have module tests at half-termly or termly intervals you can analyse student responses from these and then use this book in follow-up lessons to help students improve their skills sets. Again, students would be directed to the skills section before practising the application of those skills in the different scientific contexts (on pages 53–85).

- An alternative approach is to use the book as a means of assessing mastery of a certain skill or skills. Students would look at one of the scientific context sections and try to answer a selection of or all the questions. An analysis of their answers will then flag up skills that need further attention and the relevant skills sections can then be looked at.

Each skill presented provides links to the biology/chemistry/physics context-based activities where students can find questions to practise those skills. Since the questions on a skill are not simply written out following the explanation of a skill, this encourages a deeper level of learning about a skill before going on to the questions. It also means that the questions can be properly set in many different contexts.

Questions that allow extended writing skills to be practised are flagged 'WEA' by the various questions in the book.

The charts on pages 7–10 show the links between all the different sections. For example, following the links from a certain skill will allow you to see which activities address that skill and to think about which skills to focus on next with students.

In whatever way you decide to use the book, we hope that you and your students will find it engaging and stimulating and, above all, give students a major boost in developing their scientific skills and their confidence in using them.

Student Book

Collins

GCSE Science 9-1
Skills Booster

2nd edition

Maths in Science, Working Scientifically and Writing Extended Answers

Mark Levesley

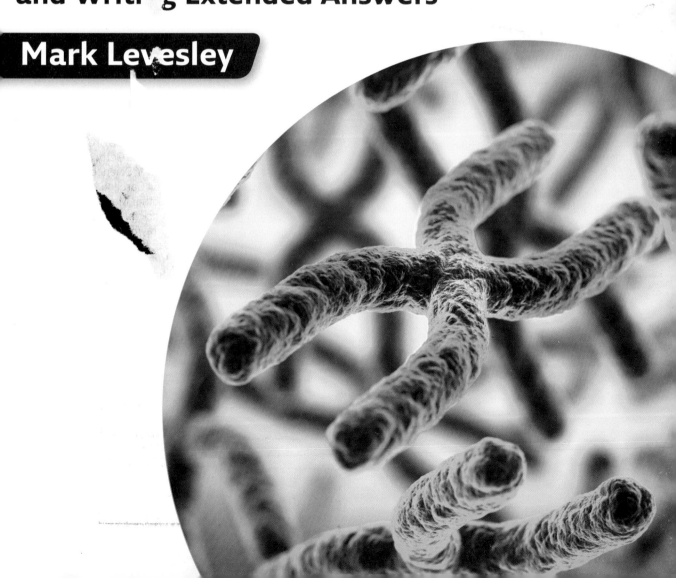

William Collins' dream of knowledge for all began with the publication of his first book in 1819. A self-educated mill worker, he not only enriched millions of lives, but also founded a flourishing publishing house. Today, staying true to this spirit, Collins books are packed with inspiration, innovation and practical expertise. They place you at the centre of a world of possibility and give you exactly what you need to explore it.

Collins. Freedom to teach

Published by Collins
An imprint of HarperCollinsPublishers
The News Building
1 London Bridge Street
SE1 9GF

Browse the complete Collins catalogue at:
www.collinseducation.com

© HarperCollinsPublishers Limited 2011

10 9 8 7 6 5 4 3 2 1

Second edition 2016

ISBN-13 978 0 00 818982 2

Mark Levesley asserts his moral right to be identified as the author of this work.

British Library Cataloguing in Publication Data
A Catalogue record for this publication is available from the British Library

Second edition commissioned by Gillian Lindsey
Copyedited and proofread by Jane Roth
Designed by Ken Vail Graphic Design
Illlustrations by Ken Vail Graphic Design,
    Phil Burrows and Beehive Illustration (Laszlo Veres)
Picture research by Caroline Green and
    Grace Glendinning
Concept design by Anna Plucinska
Cover design by Julie Martin
Production by Rachel Weaver

Printed and bound by Martins the Printers

Credits
With many thanks to: Lianne Parkin, Sheila Williams and Patricia Priest from the Department of Preventive and Social Medicine, University of Otago, New Zealand for providing the raw data from their study (page 78/P4) and Chris Pearce for his perceptive review of the manuscript.

Acknowledgements
The publishers wish to thank the following for permission to reproduce photographs. Every effort has been made to trace copyright holders and to obtain their permission for the use of copyright materials. The publishers will gladly receive any information enabling them to rectify any error or omission at the first opportunity.

Photo credits
Cover image koya979/Shutterstock; p. 14 Chris Turner/Shutterstock; p. 17 Franck Boston/Shutterstock; p. 19 Jenny Matthews/Alamy; p. 20 Swapan/Shutterstock; p. 28r AISPIX /Shutterstock; p. 29 Chuck Nacke/Alamy; p. 32c Carolina K. Smith, M.D./Shutterstock; p. 34 Wronkiew/WikiMedia Commons; p. 35 Jeremy Mayes/iStockphoto; p. 37 AFP/Getty Images; p. 38b Africa Studio/Shutterstock; p. 38cb Yuri Arcurs/Shutterstock; p. 38cc l i g h t p o e t/Shutterstock; p. 38cl Toranico/Shutterstock; p. 38ct wavebreakmedia ltd/Shutterstock; p. 38t Martin Shields /Alamy; p. 40 Four Oaks/Shutterstock; p. 41 Andrew Lambert Photography/Science Photo Library; p. 43 PHILIPPE GONTIER/EURELIOS/Science Photo Library; p. 44bc HSE; p. 44bl HSE; p. 44br HSE; p. 44tc HSE; p. 44tl HSE; p. 44tr HSE; p. 47 Diego Barbieri/Shutterstock; p. 53l Per-Anders Pettersson/Getty Images; p. 53r Judy Whitton/Shutterstock; p. 54l STEVE GSCHMEISSNER/Science Photo Library; p. 54r Mau Horng/Shutterstock; p. 55t Arnoud Quanjer/Shutterstock; p. 56 Maridav/Shutterstock; p. 57 PASCAL GOETGHELUCK/Science Photo Library; p. 58 Barcroft Media via Getty Images; p. 59 NASA; p. 60 DR M.A. ANSARY/Science Photo Library; p. 63 Caroline Green; p. 64 Carolina K. Smith, M.D./Shutterstock; p. 65 Julius Lothar Meyer; p. 66 KENCKOphotography/Shutterstock; p. Manchester Evening News; p. 70 Iain McGillivray/Shutterstock; p. 72 Charles D. Winters/Photo Researchers, Inc./Science Photo Library; p. 74t nolimitpictures/iStockphoto; p. 75 RIA Novosti/Alamy; p. 76l brejetina/Shutterstock; p. 76r ERICH SCHREMPP/Science Photo Library; p. 77l AD Singh, P Bhatnagar, B Bybel/Department of Ophthalmic Oncology, Cole Eye Institute and Department of Molecular and Functional Imaging, Cleveland Clinic Foundation, Cleveland, OH, USA; p. 77r AD Singh, P Bhatnagar, B Bybel/Department of Ophthalmic Oncology, Cole Eye Institute and Department of Molecular and Functional Imaging, Cleveland Clinic Foundation, Cleveland, OH, USA; p. 78 Courtesy of David C. Holzman; p. 79 Gary Louth/Manchester Evening News; p. 80 NASA, ESA, and M. Showalter (SETI institute); p. 81 clearviewstock/Shutterstock; p. 82 Tony Hobbs/Alamy; p. 83 David Hawgood/WikiMedia Commons

Contents

For the student

The reason why science is a compulsory subject for much of your time at school is because it teaches you how to think in ways that will help you throughout your life. Science can be divided into two parts:

* learning about what science has already discovered
* learning how that science has been discovered (scientific skills).

Sometimes we spend too much time in science learning about stuff that scientists have already discovered and so we forget to develop our own skills of scientific discovery. It is these skills that will help you to understand new technology as it becomes available, to interpret data presented on TV and the internet, and to question claims made by people and advertisers.

This book is designed to help you build your confidence in using scientific skills. There's some maths too and there's also help with using English to clearly express your scientific views and ideas.

The book starts with a section that will help you develop the skills you need. You can then follow the links from each skill to activities in the next section and so practise using your skills in different contexts.

You can also use the book the other way around, and try out some of the activities in the second section first. If you get stuck on a question, just follow its link to the relevant skills section. This will then help you develop the skills you need to answer the question.

Skills

The skills required by the GCSE specifications are explained in detail with clear, fully illustrated explanations.

WEA Some questions in GCSE exam papers require an extended piece of writing, and may be worth up to 6 marks. This 'Writing Extended Answers' icon indicates skills that will help you develop good habits when answering these sorts of questions.

Name Check! is used to identify something that has more than one term to describe it.

The meanings of all **bold words** in the text can be found in the glossary at the back of the book.

Easy cross references let you find questions to practise new skills you want to master.

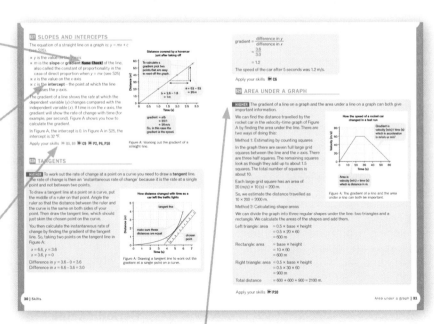

HIGHER The 'Higher' icon flags skills that will only be tested on the Higher Tier route through a GCSE course.

Skills coverage

There are two charts here to help you plan your teaching. The simple chart below lists the activities in the biology/chemistry/physics context sections and which skills are practised in them. On the next three pages the skills described in the book are listed, together with how they are linked to one another and to the Maths and Working Scientifically skills, and which activities in this book can be used to build them.

Activity	Page	Skills boosted
B1 Hairy rhinos	53	3, 5, 12, 13, 32, 45, 58, 59
B2 Salty snacks	54	3, 31, 37, 40, 41, 43, 48, 49, 58, 59
B3 Dar-wins!	55	2, 12, 17, 23, 34, 39, 40, 41, 42, 58, 59
B4 Fishy treatments	56	3, 5, 9, 11, 23, 38, 39, 40, 43, 45, 53, 58, 59
B5 Brain size	57	2, 6, 15, 16, 24, 27, 41, 42, 51, 52, 54, 55, 58, 59, 60
B6 Plants in space	59	6, 14, 18, 23, 31, 46, 48, 50, 56, 58, 59
B7 Health scare	60	12, 16, 23, 38, 39, 49, 58, 59
B8 Estimating population size	61	3, 12, 32, 44, 47, 53, 58
B9 More than a tan	62	2, 10, 13, 15, 16, 24, 27, 35, 36, 39, 40, 58, 59
B10 BMI	63	4, 12, 14, 16, 21, 32, 43, 55, 58, 59
C1 Memory of water	64	3, 6, 12, 31, 39, 43, 49, 51, 58, 59
C2 Periodic table	65	1, 2, 7, 18, 23, 34, 39, 40, 42, 55, 57, 58, 59
C3 Testing materials	66	3, 4, 5, 15, 20, 23, 32, 36, 43, 48, 56, 58, 59
C4 Chemistry and war	67	7, 16, 38, 41, 43, 47, 55, 58, 59, 60
C5 River quality	68	12, 14, 18, 36, 41, 50, 56, 58, 59
C6 Rates of reaction	69	3, 5, 23, 28, 32, 34, 40, 43, 44, 49, 58, 59
C7 Earth warming	70	9, 12, 13, 24, 27, 31, 37, 38, 58, 59
C8 Airbags	72	18, 25, 32, 36, 38, 43, 45, 47, 48, 58, 59
C9 Avogadro's big idea	73	3, 5, 26, 27, 31, 33, 39, 56, 58
C10 Cold packs	74	6, 16, 23, 40, 43, 47, 48, 51, 52, 54, 58, 59
P1 Cold fusion	75	2, 33, 36, 38, 39, 40, 46, 49, 52, 58, 59
P2 Galileo's idea	76	7, 15, 16, 19, 20, 21, 22, 27, 33, 41, 49, 51, 55, 56, 57, 58
P3 PET scans	77	3, 4, 5, 6, 31, 35, 36, 40, 45, 47, 51, 57, 58, 59, 60
P4 On the socks	78	6, 8, 12, 23, 35, 38, 43, 46, 47, 48, 58, 59, 60
P5 Sound advice	79	2, 16, 18, 38, 42, 48, 52, 53, 55, 56, 57, 58, 59
P6 Astronomical	80	3, 25, 26, 30, 34, 39, 40, 50, 51, 58, 59
P7 SPF	82	3, 6, 7, 16, 23, 37, 43, 47, 50, 54, 58, 59, 60
P8 Speed limits	83	3, 13, 14, 19, 21, 23, 37, 38, 56, 58, 59
P9 Measuring distances	84	3, 6, 15, 18, 20, 43, 45, 51, 54, 58, 59
P10 Springs	85	5, 6, 7, 18, 20, 21, 23, 24, 25, 26, 27, 29, 50, 51, 58

Biology activities: B1–B10
Chemistry activities: C1–C10
Physics activities: P1–P10

The Working Scientifically skills referred to are:

- WS1 The development of scientific thinking
- WS2 Experimental skills and strategies
- WS3 Analysis and evaluation
- WS4 Vocabulary, units, symbols an d nomenclature.

The Maths skills referred to are those in the GCSE science specifications.

	Skill	Page	Maths skills	Working Scientifically	Links to other skills in this book	Links to activities in this book
Maths in Science	S1 Decimals	11	1a	WS4	S2, S3, S4, S17	**C2**
	S2 Standard form	11	1b	WS4	S1, S18, S55, S57	B3, B5, B9, **C2, P1, P5**
	S3 Fractions, percentages and ratios	12	1c	WS3	S1, S13, S14, S18, S20, S32	B1, B2, B4, B8, **C1, C3, C6, C8, C9, P3, P6, P7, P8, P9**
	S4 Estimates: rounding	13	1d	WS4	S5, S6, S12, S58	B10, **P3**
	S5 Significant figures	14	2a	WS4	S4, S20, S50	B1, B4, **C3, C6, C9, P3, P10**
	S6 Means and ranges	14	2b	WS3	S4, S5, S14, S48	B5, B6, **C1, C10, P3, P4, P7, P9, P10**
	S7 Recording data in tables	15	2c	WS2	S8, S23, S43	B10, **C2, C4, C9, P2, P7, P10**
	S8 Frequency tables and diagrams	16	2c	WS3	S7, S9, S10, S23	**P4**
	S9 Bar charts	17	2c	WS3	S8, S10, S23, S41, S50	B4, **C7**
	S10 Charts with continuous data	18	2c	WS3	S8, S9, S23, S41	B9
	S11 Pie charts	19	2c	WS3	S23, S30	B1, B4
	S12 Estimates: sampling	19	2d	WS2	S5, S44, S50, S54, S58	B1, B3, B7, B8, B9, B10, **C1, C5, C7, P4**
	S13 Probability	20	2e	WS3	S3	B1, B9, **C7, P8**
	S14 Mean, mode, median and percentiles	21	2f	WS3	S3, S4, S16, S48	B4, B5, B6, B10, **C5, C10, P2, P4, P7, P8**
	S15 Scatter diagrams	22	2g	WS3	S16, S23, S25, S26, S29, S48, S50, S51	B5, B9, **C3, C9, P2, P9**
	S16 Correlations	23	2g	WS3	S15, S25, S43, S52	B5, B7, B9, B10, **C4, C10, P5, P7**
	S17 Orders of magnitude	23	2h	WS4	S1	B3
	S18 Symbols and conventions	24	3a	WS4	S2, S3, S31, S47, S55, S56, S57	B6, **C2, C5, C8, P5, P9, P10**

S19 Equations: changing the subject	24	3b	WS3	S20, S22, S31, S52	**P8**
S20 Equations: substituting	25	3c	WS3	S3, S5, S19, S21, S22, S32, S52	**C3, P2, P9, P10**
S21 Interconverting units	26	3d	WS3	S20, S22, S52, S55, S56, S57	B10, **P2, P8, P10**
S22 Equations: solving	26	3d	WS3	S19, S20, S21, S58	**P2**
S23 Presenting data	27	4a	WS3	S7, S8, S9, S10, S11, S15, S16, S24, S33, S52, S59	B6, B7, **C2, C3, C6, C7, C10, P4, P6, P7, P8, P9, P10**
S24 Drawing line graphs from data	28	4a	WS3	S23, S26, S28, S29, S41, S43	B9, **C7, P5, P10**
S25 Linear relationships	28	4b	WS3	S15, S16, S27	**P6, P10**
S26 Plotting variables	29	4c	WS3	S15, S24, S43	**C9, P6, P9, P10**
S27 Slopes and intercepts	30	4d	WS3	S25, S28, S29	B5, B9, **C9, P2, P6, P10**
S28 Tangents	30	4e	WS3	S24, S27, S29	**C6**
S29 Area under a graph	31	4f	WS3	S15, S24, S27, S28, S32,	**P10**
S30 Measuring angles	32	5a	WS2	S11, S44	**P6**
S31 Models	32	5b	WS1	S18, S19, S33, S34	B2, B6, **C1, C3, C6, C7, C9, P3, P8**
S32 Calculating areas, volumes and perimeters	33	5c	WS3	S3, S20, S29, S55, S57, S58	B1, B8, B10, **C3, C6, C8, P3**
S33 Development of scientific theories	34		WS1	S23, S31, S34, S35, S39, S40, S42, S43, S44, S46, S52	B3, **C9, P1, P2**
S34 Hypotheses and theories	35		WS1	S31, S33, S40	B3, **C2, C6, P6**
S35 Scientific questions	35		WS1	S33, S38, S40	B9, **P3, P4**
S36 Benefits, drawbacks and risks	36		WS1	S37, S38, S47	B9, **C3, C5, C8, P1, P3**
S37 Decisions: risks	36		WS1	S12, S36, S38, S47, S58	B2, **C7, P7, P8**
S38 Decisions: limitations and ethics	37		WS1	S35, S36, S37, S49, S53, S58	B4, B6, B7, **C4, C8, P1, P4, P5**
S39 Peer review	38		WS1	S33, S48, S49, S50, S51, S52, S53, S54	B3, B7, B9, **C1, C2, C9, P1, P6**
S40 Developing hypotheses and predictions	39		WS2	S33, S34, S35, S52	B2, B3, B4, B9, **C2, C6, C10, P1, P3, P6**

Working Scientifically (row label spanning S33–S40)

S41 Qualitative and quantitative data	40		WS2	S7, S8, S9, S10, S11, S15, S16, S23, S24, S42	B2, B5, **C5, P2**
S42 Primary and secondary data	40		WS2	S33, S41	B3, B5, **C2, P5**
S43 Variables and fair tests	41		WS2	S7, S15, S23, S24, S26, S33, S44, S45, S46, S49, S53	B2, B4, B10, **C1, C3, C6, C8, C10, P4, P7, P9**
S44 Choosing materials and methods	41		WS2	S12, S30, S33, S43, S45	B8, **C6**
S45 Trial runs	43		WS2	S33, S43, S44, S47	B1, B4, **C8, P3, P9**
S46 Controls	43		WS2	S33, S43	B2, B6, B9, **P1, P4**
S47 Safety: risks and hazards	44		WS2	S18, S36, S37, S45	B8, **C4, C8, C10, P3, P4, P7**
S48 Anomalous results and outliers	44		WS3	S6, S14, S15, S39, S51, S54	B2, B6, **C3, C8, C10, P4, P5**
S49 Repeatability and reproducibility	45		WS3	S38, S39, S43, S52, S54	B2, B7, **C1, C6, P1, P2**
S50 Accuracy, precision and uncertainty	45		WS3	S5, S9, S12, S15, S39, S51, S54	B6, **C5, C6, P7, P10**
S51 Errors in measurements	46		WS3	S9, S15, S39, S48, S50, S54	B5, **C1, C10, P2, P3, P6, P9, P10**
S52 Conclusions	46		WS3	S16, S19, S20, S21, S23, S33, S39, S40, S49, S53, S54	B5, B6, B7, B10, **C3, C10, P1, P5**
S53 Validity	47		WS3	S38, S39, S43, S52, S54	B4, B8, B9, **C3, P5**
S54 Evaluating	47		WS2, WS3	S12, S39, S48, S49, S50, S51, S52, S53, S58	B5, B9, **C3, C10, P7, P9**
S55 The SI system	48		WS4	S2, S18, S21, S32, S56, S57	B5, B10, **C2, C3, C4, C8, C9, P2, P5, P8**
S56 Compound units	49		WS4	S18, S21, S55	B6, B10, **C3, C5, C6, C9, P2, P8**
S57 Index form	49		WS4	S2, S18, S21, S32, S55	B1, B5, B10, **C2, P2, P3**
S58 Command words	50			S4, S12, S22, S32, S37, S38, S54, S59, S60	B1–10, **C1–10, P1–10**
S59 Long-answer questions	51			S23, S58, S60	B1–B7, B9, B10, **C1–C5, C7, C8, C10, P1, P3–P9**
S60 Discussion questions	52			S58, S59	B5, **C4, P3, P4, P7**

Exam Skills (vertical label spanning S58–S60)

S1 DECIMALS

If a number has no decimal point it is an **integer**.
A positive integer is called a **whole number**.

The position or 'place value' of each digit in a number is important. In an integer, the digit on the far right is in the 'ones' place. The digit in the position to the left of this is in the 'tens' place, and the next digit to the left is in the 'hundreds' place. For each move to the left, the value of the digit is 10 times bigger.

In a **decimal number** the decimal point separates whole numbers on the left from parts of whole numbers on the right. The place value of the digit just after the decimal point is 'tenths'. The place value of the next digit to the right is 'hundredths'. For each move to the right after the decimal point, the value of the digit is 10 times smaller. In 3469.425 the digit '2' is worth 2 hundredths and the digit '5' is worth 5 thousandths.

Apply your skills ▶▶ **C2**

thousands (1000s)	hundreds (100s)	tens (10s)	ones (1s)

3469

Figure A: An integer.

tenths $\left(\frac{1}{10}\right)$	hundredths $\left(\frac{1}{100}\right)$	thousandths $\left(\frac{1}{1000}\right)$

3469.425

decimal point

Figure B: A decimal number.

S2 STANDARD FORM

Standard form is a way of writing very large or very small numbers in a way that's easier to understand. To write a large number in standard form we shift the decimal point to the left until we have a single digit (1 to 9). We then write the number of places the decimal point has shifted as a power of 10 ('10' with an **index**).

e.g. $629\,000\,000\,000$ is 6.29×10^{11}

The index is 11 because we shifted the decimal point 11 places to the left. This is the same as saying 6.29 multiplied by $10 \times 10 \times 10 \times 10 \times 10 \times 10 \times 10 \times 10 \times 10 \times 10 \times 10$. For very small numbers we do the opposite and use a negative index.

e.g. $0.000\,000\,000\,000\,54$ is 5.4×10^{-13}

To add or subtract numbers in standard form, you convert all the numbers to the same power of 10. Then add or subtract the main number and leave the index as it is.

e.g. $3 \times 10^6 + 4 \times 10^7$
$= 3 \times 10^6 + 40 \times 10^6$
$= 43 \times 10^6$ or 4.3×10^7

$8 \times 10^8 - 3 \times 10^7$
$= 8 \times 10^8 - 0.3 \times 10^8 = 7.7 \times 10^8$

To divide in standard form, you divide the big numbers and subtract the indices.

e.g. $3 \times 10^6 \div 2 \times 10^5$
So, $3 \div 2 = 1.5$ and $6 - 5 = 1$
The answer is 1.5×10^1 or 15.

Continued overleaf ➔

To multiply, you multiply the main numbers together and add the indices.

e.g. $4 \times 10^{-6} \times 8 \times 10^{-5}$
So, $4 \times 8 = 32$ and $-6 + -5 = -11$.

The answer comes out as 32×10^{-11} which is 3.2×10^{-10}.

You'll usually use a calculator for these calculations. To enter a number in standard form, put in the first number, then press EXP and then enter the index.

When you write quantities in standard form, it is usual to use the base SI unit (see S55). So, for example, you should write 4.0×10^{-4} m (base unit for length) and not 4.0×10^{-1} mm.

Apply your skills ▶▶ **B3, B5, B9** ▶▶ **C2** ▶▶ **P1, P5**

S3 FRACTIONS, PERCENTAGES AND RATIOS

A **fraction** tells you how much of something there is. The 'thing' is divided into equal parts and the *total possible number* of parts goes on the bottom of the fraction. This is the **denominator**. The *actual number* of parts is put on the top. This is the **numerator**.

Scientific calculators can be used to add, subtract, multiply or divide fractions. Make sure you have a calculator that can do this and you know how to use it.

A **percentage** (%) is a fraction in which the denominator is 100. So 15% means '15 parts out of 100'. To show a fraction as a percentage, multiply it by 100. You should be able to do this with a calculator.

You might want to calculate a percentage of a number. To do this, you multiply the number by the percentage, as a fraction or decimal. For example to calculate 9% of 2000, multiply 2000 by 9/100, which is the same as 2000×0.09.

To find out how much larger (or smaller) one number is compared to another, you divide one number by the other and multiply by 100. So, for example, if we want to know what percentage 6 is of 12, we write '6 of 12' as '6/12'. Then, $6/12 = \frac{1}{2}$ or 0.5. As a percentage 0.5×100 = 50%. So 6 is 50% of 12.

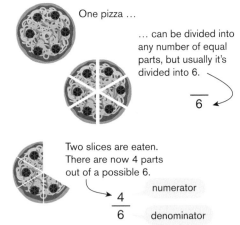

One pizza …

… can be divided into any number of equal parts, but usually it's divided into 6.
$$\overline{6}$$

Two slices are eaten. There are now 4 parts out of a possible 6.

$$\frac{4}{6}$$ numerator
denominator

Figure A: Fractions.

$$\frac{5}{8} \times 100 = 0.625 \times 100 = 62.5\%$$

To do this on a calculator:

5 ÷ 8 %

or 5 ÷ 8 × 1 0 0

Figure B: Converting fractions to percentages.

You are often asked to work out a **percentage change**. For example:

Question: What is the percentage change in mass when 2.00 g of magnesium reacts with oxygen to form 3.33 g of magnesium oxide?

Step 1: Take the number at the end of the process and subtract the original.

final mass – starting mass
3.33 g – 2.00 g = 1.33 g

Step 2: Divide the increase by the original number.

$$\frac{1.33\ g}{2.00\ g} = 0.665$$

Step 3: Multiply by 100.

% change in mass = 0.665 × 100 = 66.5%

A positive number is a percentage increase.
A negative number is a percentage decrease.

A **ratio** is another way to compare two quantities.
If Team A has 2 points and Team B has 3 points there is a '2 to 3 ratio', which is written as 2:3.

Ratios should be cancelled down to their simplest forms by dividing both sides by the same number. For example, a ratio of 3:9 is simplified to 1:3 by dividing both sides of the ':' by 3.

Apply your skills ▶▶ **B1, B2, B4, B8** ▶▶ **C1, C3, C6, C8, C9**
▶▶ **P3, P6, P7, P8, P9**

- To convert a fraction to a decimal, divide the numerator by the denominator.

$$\frac{3}{4} = 3 \div 4 = 0.75$$

- To convert a decimal to a fraction find out how many tenths, hundredths etc. there are after the decimal point and show these as a fraction.

2.4 is 2 and 4 tenths = $2\frac{4}{10} = 2\frac{2}{5}$

8.35 is 8 and 35 hundredths

$$= 8\frac{35}{100} = 8\frac{7}{20}$$

Figure C: Converting decimals and fractions.

S4 ESTIMATES: ROUNDING

It is often useful to **estimate** the result of a calculation. This means doing a rough calculation. Estimating:
- saves time
- helps you to focus on the important parts of what you are doing.

One way of estimating is to **round** figures up or down before doing a calculation. For example, if you wanted to know if £10 was enough to buy 3 coffees at £2.89 each you could say 'Well £2.89 is roughly £3 and 3 × £3 is £9, so I have enough'. That's a much easier calculation than 3 × £2.89 and you don't need to know the exact cost. Or imagine you want to know the approximate area of a room which is 3.25 m by 1.96 m. You can round these values to calculate 3 × 2 = 6 m².

Scientists use the symbol ~ to show that a number is roughly right. For example, ~5 cm means 'roughly 5 centimetres'.

Apply your skills ▶▶ **B10** ▶▶ **P3**

Small coffee £1.11
Medium coffee £2.07
Large coffee £2.89

Three large coffees please.

Hang on, can we afford that?

Figure A: Estimating saves time.

S5 SIGNIFICANT FIGURES

Significant figures are the digits that show a value's **accuracy**. For example, a town's population is 10675 people. To two significant figures this is 11000 (only two figures show the amount, the rest are zeros). 11000 is a less accurate figure than 10675.

Less accurate measuring devices produce values with fewer significant figures, compared with more accurate devices.

When doing calculations, do not give your answer to more significant figures than the least accurate value that you started with. For example:

Question: Use the equation $KE = \frac{1}{2} \times m \times v^2$ to calculate the kinetic energy stored by a car with mass 2100 kg and speed 3.7 m/s.

$KE = \frac{1}{2} \times m \times v^2$

$KE = \frac{1}{2} \times 2100 \times (3.7 \times 3.7)$

$KE = 28\,749$ J ✗ (too many significant figures)

$KE = 29\,000$ J ✔ (number of significant figures matches the given values)

Apply your skills ▶▶ **B1, B4** ▶▶ **C3, C6, C9** ▶▶ **P3, P10**

289687	To show this to 3 significant figures	0.0034523
289687	Count three digits from the left (starting from the first digit that is not a 0)	0.0034523
289687	Look at the digit after the significant figures. If it's 5 or greater, then increase the last significant figure by 1.	0.0034523
290000	Convert all the numbers to the right of your significant figures to 0.	0.0034500
290000 (3 sf)	State the number of significant figures in brackets. This step is not usually necessary.	0.00345 (3 sf)

Figure A: Changing significant figures.

S6 MEANS AND RANGES

Repeated measurements usually vary. In maths, **range** of the measurements is the calculated difference between the highest and lowest values, often ignoring any **anomalous results** (**outliers**) that can be explained (see S48). The narrower the range of repeated measurements, the more sure you can be that they are correct. In science, a range also refers to a statement saying what the highest and lowest values are.

To determine the best value from a set of values, we can calculate an average called a **mean**:
- add up all your measurements
- divide by the number of measurements you took
- change the answer to the same number of **significant figures** as your original data.

The more measurements used in the calculation, the closer the mean will be to the true value, but the longer it takes. We can sometimes also get a better mean value by leaving out any outliers, but only when we are confident we can explain why they occurred.

How long do I take to zorb down this track?

Figure A: The mean time has been calculated in the example on the next page.

Test number	Time taken to zorb down the track (s)
1	24.0
2	24.6
3	24.5
4	82.7
5	24.5
6	24.4

Calculating a mean:

There is one outlier in the data in the table – Test 4. This was caused by forgetting to stop the stopwatch. We ignore it, otherwise it will give us a poor estimate of the time taken.

Total = 122.6

Number of tests = 5

Mean = $\frac{122.6}{5}$ = 24.4 s

Apply your skills ⏩ **B5, B6** ⏩ **C1, C10** ⏩ **P3, P4, P7, P9, P10**

S7 RECORDING DATA IN TABLES

Tables are used to record the values of **variables**. Tables do not contain information about **control variables** (see S43) or how variables were measured. There is a standard way of setting out a table of data, as shown in Figure A.

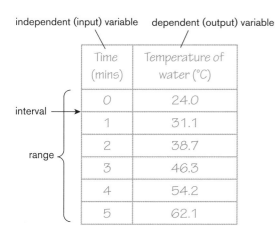

Figure A: Temperature increase of 300 cm³ of water heated by burning 30 cm³ of oak wood.

If you design a table when planning an investigation, it helps you think about:
- your choice of **input variable** or **independent variable** **Name Check!** (the one you will change in the investigation) and **output variable** or **dependent variable** **Name Check!** (the one you will measure in the investigation)
- the **range** of measurements to use
- the intervals to have between measurements
- how to measure the dependent variable.

Tables also allow you to present data in different orders. This helps you to look for patterns.

Apply your skills ⏩ **B10** ⏩ **C2, C4, C9** ⏩ **P2, P7, P10**

S8 FREQUENCY TABLES AND DIAGRAMS

The **frequency** of something is how often it occurs. We can record this in a **frequency table** (or **tally chart**) **Name Check!** , which has three columns. The things you are counting go in the first column. An example is shown in Figure A.

The second column is the tally. To complete it, look through your data and each time you find an item in the first column, place a vertical line in its tally. When you have four vertical lines, the next line is drawn diagonally through the vertical lines to show a count of '5'.

Currency symbols	Tally	Frequency
£	IIII	4
€	~~IIII~~ II	7
$	~~IIII~~	5

Data

£	$	$	€
€	€	€	$
£	£	€	€
$	€	$	£

Figure A: A frequency table, or tally chart, showing the number of occurrences of currency symbols in the data.

When you have looked at all your data, use the scores in the tally column to work out the values in the frequency column.

Sometimes, the first column will include groups. You complete the chart in the same way but this time you need to look to see which group each piece of your data belongs in. This is shown in Figure B.

Height groups	Tally	Frequency
1.50–1.59 m	II	2
1.60–1.69 m	III	3
1.70–1.79 m	~~IIII~~	5
1.80–1.89 m	IIII	4

Heights in Year 11

1.76 m	1.70 m
1.56 m	1.87 m
1.60 m	1.67 m
1.75 m	1.83 m
1.61 m	1.84 m
1.82 m	1.77 m
1.72 m	1.57 m

Figure B: A frequency table for grouped data.

Histograms (see S10) and many **bar charts** (see S9) show frequencies visually. These are called **frequency diagrams**.

Apply your skills ⏩ **P4**

Bar charts are often used to present data in which the independent variable is either **qualitative** (**categoric**) or **discrete**. The dependent variable (the one that you measure) is **quantitative**. (See S41 for an explanation of the difference between qualitative and quantitative data.)

Generally, the independent variable is plotted on the horizontal axis (*x*-axis). The dependent variable is on the vertical axis (*y*-axis). It can be done the other way around but the bars always come out of whichever axis the independent variable is on.

Error bars may be added to show the range of measurements used to calculate average measurements and indicate the degree of **uncertainty**. Shorter error bars show measurements of greater **precision**.

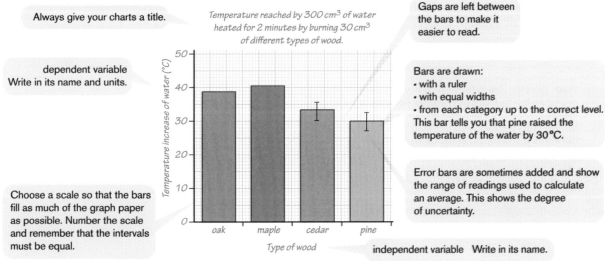

Always give your charts a title.

Gaps are left between the bars to make it easier to read.

dependent variable Write in its name and units.

Bars are drawn:
• with a ruler
• with equal widths
• from each category up to the correct level. This bar tells you that pine raised the temperature of the water by 30 °C.

Error bars are sometimes added and show the range of readings used to calculate an average. This shows the degree of uncertainty.

Choose a scale so that the bars fill as much of the graph paper as possible. Number the scale and remember that the intervals must be equal.

independent variable Write in its name.

Figure A: A bar chart.

If you want to show how several things in different groups change, you can group the bars together.

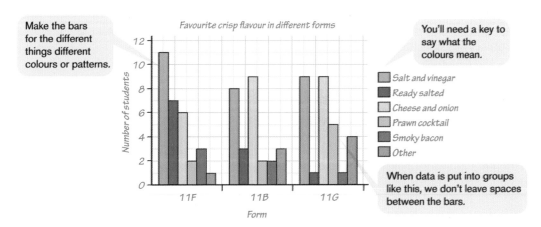

Make the bars for the different things different colours or patterns.

You'll need a key to say what the colours mean.

When data is put into groups like this, we don't leave spaces between the bars.

Apply your skills ▶▶ **B4** ▶▶ **C7**

Figure B: A bar chart with bars shown in groups.

Continuous data, as well as **discrete data** (see Bar charts, S9) can be plotted using bars on a chart. To do this, we divide the data into groups. Figure A shows an example. It is a bar chart and a **frequency diagram** (because frequency is on the vertical axis).

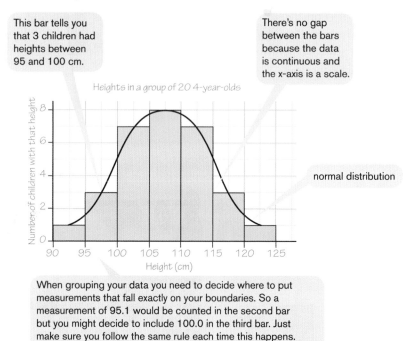

This bar tells you that 3 children had heights between 95 and 100 cm.

There's no gap between the bars because the data is continuous and the x-axis is a scale.

Heights in a group of 20 4-year-olds

normal distribution

When grouping your data you need to decide where to put measurements that fall exactly on your boundaries. So a measurement of 95.1 would be counted in the second bar but you might decide to include 100.0 in the third bar. Just make sure you follow the same rule each time this happens.

Figure A: A bar chart of grouped continuous data.

When you plot a frequency diagram for grouped continuous data it often forms a 'bell shape' (as shown by the curve drawn on Figure A). This shape is called a 'normal distribution' because it is how we expect continuous data to be spread out.

A **histogram** is a special kind of bar chart. The width of each bar in a histogram must match the range of that group of data (the difference between the highest and lowest values in the group). In Figure A all bars are the same width, corresponding to a range of 5 cm. But it is possible for groups to have different ranges and so in a histogram the bars would have different widths. The frequency is then shown by the *area* of each bar. A histogram has 'frequency density' plotted on the vertical axis. This is worked out for each group by dividing the frequency of that group by its range.

Apply your skills ▶▶ **B9**

Pie charts are used to compare the contributions made by different categories to a whole. A pie chart is a circle and so has 360°. These 360° are divided up into the same proportions as the different categories.

Favourite flavour of crisps	Number of students in class 11F
cheese & onion	6
other	1
prawn cocktail	2
ready salted	7
salt & vinegar	11
smoky bacon	3

Step 1: Add up the total number.
total is 30

Step 2: Divide 360° by the total number.
360/30 = 12 (so each student is represented by 12° of the circle)

Step 3: Multiply the number in each category by your answer to step 2.
e.g. 6 × 12 = 72

Step 4: Add a third column to the table and write the angles in there.

Include a title.

Neatly label your categories.

Favourite crisps in class IIF

smoky bacon

cheese & onion

Use a protractor to measure angles.

other

prawn cocktail

salt and vinegar

ready salted

Use a ruler to draw the lines.

Use a compass or similar to draw a circle.

Figure A: Drawing a pie chart.

Apply your skills ▶▶ **B1, B4**

S12 ESTIMATES: SAMPLING AND BIAS

Data can be collected in a variety of ways. For example, it may be collected by doing an experiment or using a survey. In a survey you count the number of things that are of interest, without changing the conditions in which the things are found.

It often takes far too long to count things (for example, the number of dandelions on playing fields). So scientists **estimate** the numbers – they use a rough calculation based on a **sample**.

For example, to work out the number of red spots in Figure A you count the spots in a small sample area. Then calculate how many times bigger the whole area is compared with the sample area. Multiplying this result by the number of red spots counted gives an estimate of the total number of spots in the whole area.

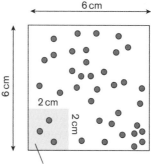

1 Choose a sample area.
2 Count spots in sample area (= 3)
3 Work out size of sample area.
   Area = length × width (= 2 × 2 = 4 cm²)
4 Work out size of whole area (= 6 × 6 = 36 cm²)
5 Calculate how many times bigger whole area is compared with sample (= 36 ÷ 4 = 9)
6 The number of spots in the whole area will be about 9 times more (= 9 × 3 = 27)
7 We estimate that there are ~27 spots in the whole area.

Figure A: By counting the spots in the small area and working out how many times that would be repeated, you can estimate how many red spots there are in total.

Continued overleaf →

Bias is when evidence is distorted in one direction. Figure B will help you to understand this.

When you make estimates using samples it is quite easy to be biased. Look at Figure C. If you estimate the number of spots in the whole drawing using square A3, it comes out as 27 spots. If you use square A2, the estimate is 18 spots and with square C3 it is 72. The actual number is 38. All our estimates are biased – two have a bias below the real value and one has a bias above it.

You get better estimates, and reduce the effects of bias, if you take more samples. However, if you take too many samples it will take too long and defeat the object of taking samples! So for the big square, you might decide to estimate using three or four smaller sample squares from the grid. But how do you pick the squares to use?

For the best estimate, you need to choose samples at **random**. Doing this means that there is no bias caused by the experimenter because the experimenter does not make the choices. So to sample a field, a biologist would divide the field into a grid (as shown in Figure C) and then use a calculator or dice to generate random numbers to choose the squares to use.

Apply your skills ▶▶ B1, B3, B7, B8, B9, B10 ▶▶ C1, C5, C7 ▶▶ P4

Although the gun was aimed at the centre of the target, the bullets all ended up in the upper left of the target. There is bias. This might be due to the gun or to the person who is firing it.

Figure B: A way of thinking about bias.

This is square A3 (row A, column 3).

Figure C: Small sample sizes can cause bias.

S13 PROBABILITY

There are six sides on a die, with a different number on each side. The die is equally likely to land showing any one of the six numbers. So if you throw the die six times, you should get each number once. This doesn't always happen because throwing a die is a **random** process and so we talk about the chance or **probability** of a number being thrown. In the case of a die, the probability of throwing a particular number is 1 in 6.

A probability can be written as a fraction, a decimal or a percentage.

Most commonly, probabilities are written as decimals on a scale of 0–1 (where 1 means it's certain that something will happen). The probabilities of all the possible outcomes always add up to 1. So, it follows that if there is a 0.17 chance of throwing a certain number on a die, there is a 1 − 0.17 = 0.83 chance of not throwing that number.

Apply your skills ▶▶ B1, B9 ▶▶ C7 ▶▶ P8

There is a 1 in 6 chance of throwing a five. This can be written as a probability, which may be shown as a fraction, percentage, or decimal of 1 or less:

$$\frac{1}{6} \text{ or } 0.17 \text{ or } 17\%$$

Figure A: Ways of writing a probability.

S14 MEAN, MODE, MEDIAN AND PERCENTILES

An **average** is a single item used to represent all the other items in a set of data. There are three different types of average that you need to know about.

* The **mode** is the most common item in a set of data. It is the only average that can be used with items that are not numbers. See Figure A.
* The **median** is the middle value of a set of values, written in order. If there are two values in the middle then you add them together and divide by two. See Figure B.
* A **mean** is an average that takes all the values in a set of data into account. You add up all the values and divide by the number of values.

Means are affected by **outliers** (values far away from all the other values) but medians are not affected. See Figure C.

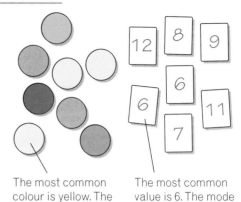

The most common colour is yellow. The mode is yellow.

The most common value is 6. The mode is 6.

Figure A: Finding a mode.

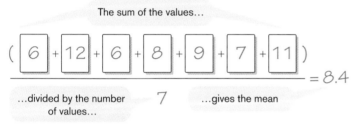

$$\underbrace{\left(\boxed{6} + \boxed{12} + \boxed{6} + \boxed{8} + \boxed{9} + \boxed{7} + \boxed{11}\right)}_{\text{The sum of the values...}}$$

...divided by the number of values...   7   ...gives the mean

$= 8.4$

Figure C: Finding a mean.

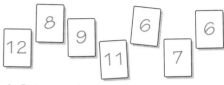

1 Put your values in order.

| 6 | 6 | 7 | 8 | 9 | 11 | 12 |

2 Count in from either end, one value at a time, until you are left with one or two values in the middle. In this case it is 8. The median is 8.

Figure B: Finding a median.

HIGHER A **percentile** is a value at which a certain percentage of some data is cut off. So the 16th percentile cuts off the first 16% of the values. In other words, 16% of the data have values that are equal to or below that value. If you are at the 16th percentile for height in your school, it would mean than 16% of students are your height or shorter.

A **decile** is the same idea but the data is divided into 10 equal parts (rather than 100). The 1st decile is the same as the 10th percentile.

A **quartile** is the same idea again but the data is divided into 4 parts. The **median** is the middle value of a whole data set, splitting a data set into 2 parts.

If n is the number of values in the data set, the median is the $0.5 \times (n + 1)$th value

The lower quartile is the middle value of the lower half of the data set, so the $0.25 \times (n + 1)$th value

The upper quartile is the middle value of the upper half of the data set, so the $= 0.75 \times (n + 1)$ th value

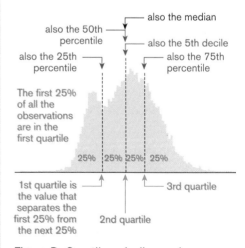

Figure D: Quartiles, deciles and percentiles.

Apply your skills ⏩ **B4, B5, B6, B10** ⏩ **C5, C10** ⏩ **P2, P4, P7, P8**

Scatter diagrams (also called **scatter graphs**, or scattergrams or scatter plots) **Name Check!** are used when you want to find a **correlation** between two variables.

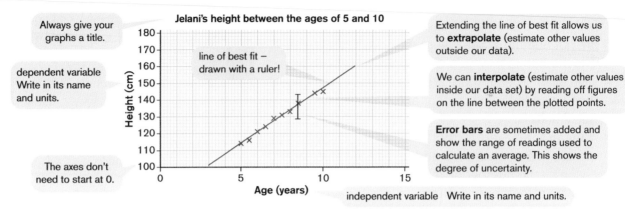

Always give your graphs a title.

dependent variable Write in its name and units.

The axes don't need to start at 0.

line of best fit – drawn with a ruler!

Extending the line of best fit allows us to **extrapolate** (estimate other values outside our data).

We can **interpolate** (estimate other values inside our data set) by reading off figures on the line between the plotted points.

Error bars are sometimes added and show the range of readings used to calculate an average. This shows the degree of uncertainty.

independent variable   Write in its name and units.

A **line of best fit** is often drawn through the points on a scatter graph. The line goes through the middle of the points, so that about half the points are on either side of it. Note that the line of best fit does not necessarily have to go through the origin (zero). You ignore any **anomalous data** (**outliers**) when drawing a line of best fit.

Figure A: A line of best fit.

For some data a **curve of best fit** is used. To draw a curve of best fit, place your paper on the desk so that the curve of points is shaped like an arch as you look at it. Then put your elbow on the desk and practise moving your hand in an arc, so that your pencil goes through the points, but don't put the pencil on the paper. You may need to experiment with the exact positioning of the paper. When you've practised a few times, put the pencil on the paper and draw a smooth curve.

For some data you will need a combination of lines of best fit and curves of best fit.

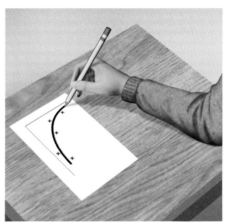

Figure B: Drawing a curve of best fit.

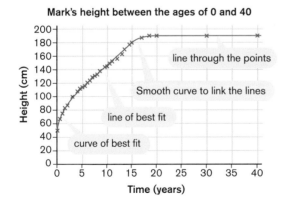

line through the points

Smooth curve to link the lines

line of best fit

curve of best fit

Figure C: Lines and curves of best fit can often be used together.

Apply your skills ▶ **B5, B9** ▶ **C3, C9** ▶ **P2, P9**

S16 CORRELATIONS

In many science experiments you want to observe how the independent (input) variable affects the dependent (output) variable. Since you want to see only the effect of the independent variable, all other variables need to be controlled (see S43).

A **correlation** is a link between changes in the independent variable and changes in the dependent variable. If a dependent variable shows a steady change when you steadily change the independent variable, there is a correlation.

A correlation can be strong or weak and positive or negative. A good way of showing a correlation is to use a **scatter diagram (scatter graph)**. **Name Check!**

A correlation may be:
* **causal** – when the independent variable (the cause) directly affects the dependent variable (the effect)
* **due to association** – when both variables are affected by another variable
* **due to chance**.

A causal correlation is more likely to be believed if there is a hypothesis that explains *how* it happens.

Apply your skills ⏩ **B5, B7, B9, B10** ⏩ **C4, C10** ⏩ **P5, P7**

No correlation

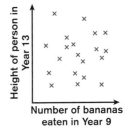

Weak positive correlation

There is a correlation but this is probably due to chance!

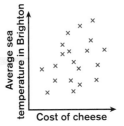

Strong positive correlation

The risk of having heart disease and the amount of cholesterol may both be being increased by another variable (e.g. lack of exercise).

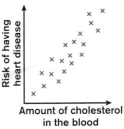

Strong negative correlation

Figure A: Scatter graphs are good for showing correlations.

S17 ORDERS OF MAGNITUDE

A division or a multiplication by 10 is 'an order of magnitude'. We use orders of magnitude to compare very small and very big things, when the approximate difference between things is more important than an accurate calculated difference.

For example, the table shows the heights of some objects that we'd like to compare.

We can say that:
* The tower in London is an order of magnitude taller than the basketball hoop.
* The Eiffel Tower is an order of magnitude taller than the tower in London.
* The Eiffel Tower is two orders of magnitude taller than the basketball hoop.
* The basketball hoop is two orders of magnitude lower than the Eiffel Tower.

Apply your skills ⏩ **B3**

Object	Height	Rounded to 1 significant figure
Basketball hoop	3.048 m	3 m
Tallest tower at Tower of London	27 m	30 m
Eiffel Tower	324 m	300 m

S18 SYMBOLS AND CONVENTIONS

A **symbol** is a way of representing something by a shape. A **convention** is a certain way of doing things. Scientists use conventions and symbols because they can:

- ⚔ speed up writing things down
- ⚔ make things look clearer
- ⚔ be understood all over the world no matter what language is spoken.

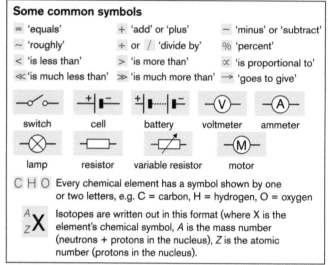

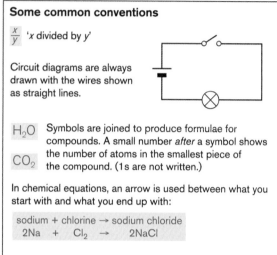

Figure A: Some common symbols and conventions used in science.

Apply your skills ▶▶ B6 ▶▶ C2, C5, C8 ▶▶ P5, P9, P10

S19 EQUATIONS: CHANGING THE SUBJECT

An equation is a way of saying that two things are equal. For example:

$$5 = 2 + 3$$

In algebra we use letters to represent quantities. An equation in which this is done is an **algebraic equation**. For example:

$$5 = x + 2$$

The **subject** of the equation is the *single* number or letter on one side of the equals sign that everything on the other side is equal to. So '5' is the subject of the equation above.

We can change the subject by performing the same actions on each side of the equals sign. If we want to remove the '2' from the right-hand side of the equation above, we need to subtract 2 from each side:

$$5 - 2 = x + 2 - 2$$
$$5 - 2 = x$$
$$3 = x$$

The algebraic equation relating velocity (v), time (t) and distance (d) is:

$$v = \frac{d}{t}$$

To make d the subject, we multiply both sides by t:

$$v \times t = \frac{d}{\cancel{t}} \times \cancel{t}$$

So, cancelling, we get:

$$v \times t = d$$

To make t the subject of $v = \dfrac{d}{t}$, we divide both sides by d and cancel:

$$\frac{v}{d} = \frac{d}{t \times d} \quad \text{so} \quad \frac{v}{d} = \frac{1}{t}$$

To turn $1/t$ into t, we can turn it the other way up. So, we also need to turn the left-hand side of the equation the other way up:

$$\frac{d}{v} = t$$

The algebraic equation relating electrical power (P), current (I) and resistance (R) is:

$$P = I^2 \times R$$

To make I the subject, we need to first get I^2 on its own and then find the square root (because the square root is the opposite of a square number).

$$\frac{P}{R} = \frac{I^2 \times R}{R} \quad \text{so} \quad \frac{P}{R} = I^2 \quad \text{and} \quad \sqrt{\frac{P}{R}} = I$$

A list of the equations you need to use is on pages 127–128.

Apply your skills ▶▶ **P8**

S20 EQUATIONS: SUBSTITUTING

An equation in which letters represent quantities is an **algebraic equation**. For example:

$$KE = \tfrac{1}{2} \times m \times v^2$$

Here, KE represents 'kinetic energy', m represents 'mass' and v represents 'speed'. If we are given the mass and the speed we can **substitute** those values into the equation to calculate the kinetic energy. For example:

Question: Calculate the kinetic energy of a car of mass 2000 kg travelling at 25 m/s.

$$KE = \tfrac{1}{2} \times m \times v^2$$

$$KE = \tfrac{1}{2} \times 2000 \text{ kg} \times (25 \text{ m/s})^2$$

$$KE = \tfrac{1}{2} \times 2000 \times (25 \times 25)$$

$$KE = \tfrac{1}{2} \times 2000 \times 625$$

$$KE = \tfrac{1}{2} \times 1\,250\,000$$

$$KE = 625\,000 \text{ J}$$

The units also get calculated. In the example the units come out as something that looks complicated: $\text{kg} \times \text{m}^2/\text{s}^2$. But this is the same as the joule (J).

Before you do a substitution:

⚊ check that you know what all the letters represent

⚊ check that the values have the right units for the equation. You may need to interconvert the units (see S21).

Apply your skills ▶▶ **C3** ▶▶ **P2, P9, P10**

S21 INTERCONVERTING UNITS

We often need to change the units of measured values.

Interconverting between larger and smaller units in the **SI system** is easy because you always use multiples of ten. For example, to convert 5.5 km into mm, we would start by saying:

1 km = 1000 m and so 5.5 km = 5.5 × 1000 = 5500 m. Then:

1 m = 1000 mm so 5500 m = 5500 × 1000 = 5 500 000 mm.

When converting between units in **index form** you need to remember to take account of the unit's index power. For example, in 1 m there are 100 cm but in 1 m^3 there are 100 cm × 100 cm × 100 cm = 1 000 000 cm^3.

We often need to interconvert units to get values into the form needed by an equation. For example, the equation below relates kinetic energy, mass and speed. All values substituted in the equation must be in SI units, so mass in kg and speed in m/s.

$KE = \frac{1}{2} \times m \times v^2$

If we were given the speed in kilometres per hour we would have to convert it to metres per second. When you interconvert **compound units** (or **compound measures** **Name Check!**) you convert each unit separately. So:

we need m/s

$90 \text{ km/h} = 90 \times 1000 \text{ m/h}$ start by converting km to m

$= 90\,000 \text{ m/h}$

$= 90\,000/60 \text{ m/min}$ convert hours into minutes

$= 1500 \text{ m/min}$

$= 1500/60 \text{ m/s}$ convert minutes into seconds

$= 25 \text{ m/s}$

Figure A: Converting compound units.

Apply your skills ▶ B10 ▶▶ P2, P8, P10

S22 EQUATIONS: SOLVING

To solve an equation we:
- write out the equation
- change the subject, if needed (see S19)
- check that the units for the values are correct for the equation and convert them if necessary (see S21)
- substitute the known values (see S20)
- calculate the result.

Apply your skills ▶▶ P2

The table shows the functions of different charts, graphs and diagrams.

Presentation type		Independent variable	Dependent variable	Used ...	Skill
Table		qualitative or quantitative	qualitative or quantitative	to record data and put it in order	S7
Bar chart		qualitative or quantitative, discrete	quantitative	to compare differences between groups	S9
Continuous data bar chart / histogram		quantitative, continuous (grouped)	quantitative (usually a frequency or frequency density)	to show how the number of times something occurs is distributed	S10
Line graph		quantitative, continuous (usually time)	quantitative	mainly to show how something changes with time	S24
Scatter graph		quantitative	quantitative	to find relationships (correlations) between variables	S15, S16
Pie chart		to compare the contributions of things to a whole			S11
Flow chart		to show how one part of a process follows another			S33, S59

Apply your skills ▶▶ **B6, B7** ▶▶ **C2, C3, C6, C7, C10** ▶▶ **P4, P6, P7, P8, P9, P10**

S24 DRAWING LINE GRAPHS FROM DATA

Line graphs are used to present data when both the independent variable and the dependent variable are in the form of **quantitative data**. They are used to show how one variable changes with another. The independent variable is often time.

The independent variable is plotted on the horizontal axis (*x*-axis). The dependent variable is plotted on the vertical axis (*y*-axis).

If you need to plot more than one line on the same graph, it's a good idea to use ×s to plot one set of points and +s or circles to plot the other.

Always give your graphs a title.

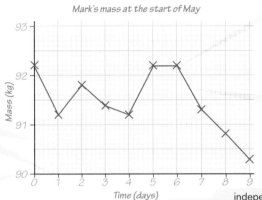

Plot each point, in pencil, with a neat ×. Then connect the points with a straight line, using a ruler.

dependent variable
Write in its name and units.

Choose axis scales so that the points fill as much of the graph paper as possible. Number the scales and remember that the intervals must be equal.

Line graphs can be used to **estimate** new values. We estimate that after 6.5 days Mark had a mass of 91.75 kg.

The axes don't need to start at 0.

independent variable
Write in its name and units.

Apply your skills ▶▶ **B9** ▶▶ **C7** ▶▶ **P5, P10**

Figure A: A line graph.

S25 LINEAR RELATIONSHIPS

Lines on graphs can be used to:
- estimate other values within your data range (**interpolation**)
- estimate other values outside your data range (**extrapolation**)
- calculate the **gradient** of a line (the resulting units will be the vertical axis units divided by the horizontal axis units)
- see if two variables are in a **linear relationship**.

Two variables have a linear relationship if, as one variable changes, the other changes so that the graph of one against the other is a straight line.

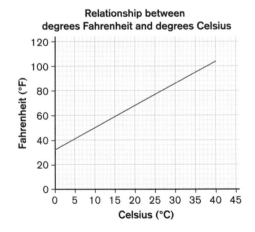

Figure A: The linear relationship between the Celsius temperature scale and the Fahrenheit scale.

If the line for a linear relationship goes through the origin, then the two variables are **directly proportional**. This means that when one changes, the other changes by the same percentage. We say that variable y is directly proportional to variable x (which is often written as $y \propto x$). The **constant of proportionality** is the amount we need to multiply an x-axis value to equal the y-axis value on the line. This constant often has the symbol 'm'. Then, to calculate values for y on a line that shows direct proportion, we can use the formula: $y = mx$.

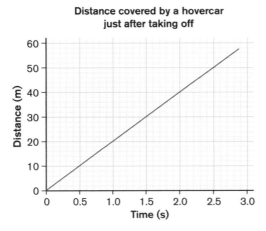

Figure B: A straight line going through the origin shows direct proportion.

If the line does not go through the origin, as in Figure A, the formula becomes $y = mx + c$, where c is the value at which the line crosses the y-axis.

HIGHER If variables are **inversely proportional**, $y \propto 1/x$ and so $y = m/x$. A graph of y against x showing inverse proportion will be a curve. To obtain a straight line, you need to plot y against $1/x$.

Apply your skills ▶▶ **P6, P10**

S26 PLOTTING VARIABLES

When plotting **variables** on charts and graphs, the **independent variable** (the one that is changed in an experiment) is plotted on the horizontal axis (x-axis). The **dependent variable** (the one that depends on the independent variable, and is measured in an experiment) is plotted on the vertical axis (y-axis).

See S15 and S24 for details on how to draw graphs.

Apply your skills ▶▶ **C9** ▶▶ **P6, P9, P10**

S27 SLOPES AND INTERCEPTS

The equation of a straight line on a graph is: $y = mx + c$ (see S25).

- ✖ y is the value on the y-axis
- ✖ m is the **slope** or **gradient** **Name Check!** of the line, also called the constant of proportionality in the case of direct proportion when $y = mx$ (see S25)
- ✖ x is the value on the x-axis
- ✖ c is the **intercept** – the point at which the line crosses the y-axis.

The gradient of a line shows the *rate* at which the dependent variable (y) changes compared with the independent variable (x). If time is on the x-axis, the gradient will show the rate of change with time (for example, per second). Figure A shows you how to calculate the gradient.

In Figure A, the intercept is 0. In Figure A in S25, the intercept is 32 °F.

Apply your skills ▶▶ **B5, B9** ▶▶ **C9** ▶▶ **P2, P6, P10**

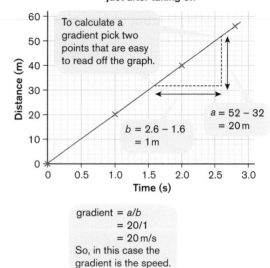

Distance covered by a hovercar just after taking off

To calculate a gradient pick two points that are easy to read off the graph.

$a = 52 - 32$
$= 20\,m$

$b = 2.6 - 1.6$
$= 1\,m$

gradient = a/b
$= 20/1$
$= 20\,m/s$
So, in this case the gradient is the speed.

Figure A: Working out the gradient of a straight line.

S28 TANGENTS

HIGHER To work out the rate of change at a point on a curve you need to draw a **tangent** line. The rate of change is then an 'instantaneous rate of change' because it is the rate at a single point and not between two points.

To draw a tangent line at a point on a curve, put the middle of a ruler on that point. Angle the ruler so that the distance between the ruler and the curve is the same on both sides of your point. Then draw the tangent line, which should just skim the chosen point on the curve.

You then calculate the instantaneous rate of change by finding the gradient of the tangent line. So, taking two points on the tangent line in Figure A:

$x = 6.6, y = 3.6$
$x = 3.6, y = 0$

Difference in $y = 3.6 - 0 = 3.6$
Difference in $x = 6.6 - 3.6 = 3.0$

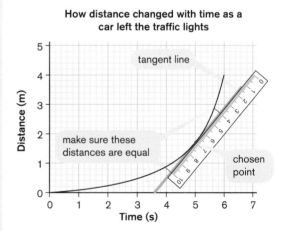

How distance changed with time as a car left the traffic lights

tangent line

make sure these distances are equal

chosen point

Figure A: Drawing a tangent line to work out the gradient at a single point on a curve.

$$\text{gradient} = \frac{\text{difference in } y}{\text{difference in } x}$$

$$= \frac{3.6}{3.0}$$

$$= 1.2$$

The speed of the car after 5 seconds was 1.2 m/s.

Apply your skills ▶ C6

S29 AREA UNDER A GRAPH

HIGHER The gradient of a line on a graph and the area under a line on a graph can both give important information.

We can find the distance travelled by the rocket car in the velocity–time graph of Figure A by finding the area under the line. There are two ways of doing this:

Method 1: Estimating by counting squares

In the graph there are seven full large grid squares between the line and the x-axis. There are three half squares. The remaining squares look as though they add up to about 1.5 squares. The total number of squares is about 10.

Each large grid square has an area of 20 (m/s) × 10 (s) = 200 m.

So, we estimate the distance travelled as 10 × 200 = 2000 m.

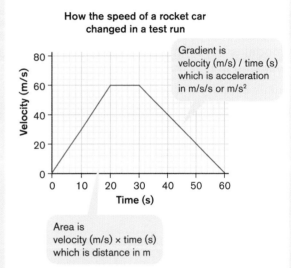

Figure A: The gradient of a line and the area under a line can both be important.

Method 2: Calculating shape areas

We can divide the graph into three regular shapes under the line: two triangles and a rectangle. We calculate the areas of the shapes and add them.

Left triangle: area	= 0.5 × base × height
	= 0.5 × 20 × 60
	= 600 m
Rectangle: area	= base × height
	= 10 × 60
	= 600 m
Right triangle: area	= 0.5 × base × height
	= 0.5 × 30 × 60
	= 900 m
Total distance	= 600 + 600 + 900 = 2100 m.

Apply your skills ▶ P10

S30 MEASURING ANGLES

Angles are measured in degrees, which have the symbol °. A right angle is 90°, a semi-circle has 180° and a full circle has 360°.

Angles can be measured using a protractor.

Protractors have two scales on them. Be careful which one you use. Remember that a right angle is 90° and so angles with a sharp point (acute angles) must be less than 90°. Obtuse angles must be greater than 90°.

This line (*not* the edge of the protractor) goes onto one of the lines of the angle you are measuring.

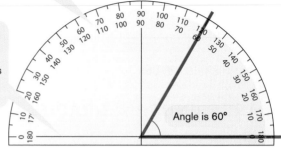

Angle is 60°

Apply your skills ▶▶ **P6**

Figure A: Using a protractor.

S31 MODELS

A **model** is anything that represents a thing or a process in a way that makes it easier for us to understand. Models usually simplify the real nature of something.

Models can be physical (you can touch them) or abstract (they are ways of thinking about things). Models can be three-dimensional (3D) or two-dimensional (2D). Circuit diagrams are abstract 2D models (see S18).

A chemical equation is an abstract model that shows what happens in a reaction.

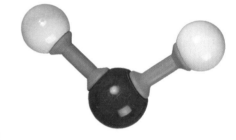

Figure A: This 3D physical model represents the atoms in a molecule of water.

hydrogen + oxygen → water

What you start with   The arrow means 'goes to give'   What you end up with

Models can also be used to think up **hypotheses** and make **predictions**. Models can be tested by observations or experiments.

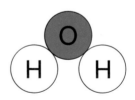

Figure B: This is a 2D model of a water molecule.

Mathematical equations are abstract models. You use the model to calculate quantities. You can test the model by comparing calculations with real data. For example, you can calculate voltage using information about current and resistance, using the equation below. You can then set up a circuit and measure the voltage. Comparing that with the calculated value allows you to test the model.

$V = IR$ This could be written $V = I \times R$ but the $\times$ is often left out.

$V = 0.1\,\text{A} \times 50\,\Omega$

$\quad = 5\,\text{V}$ If we know the values of I (the current) and R (the resistance), we can calculate V.

Apply your skills ▶▶ **B2, B6** ▶▶ **C1, C3, C6, C7, C9** ▶▶ **P3, P8**

Figure A shows the ways in which the **areas**, **volumes** and **perimeters** of different shapes are calculated.

b = base
d = diameter
h = height
l = length
r = radius
w = width
π = pi

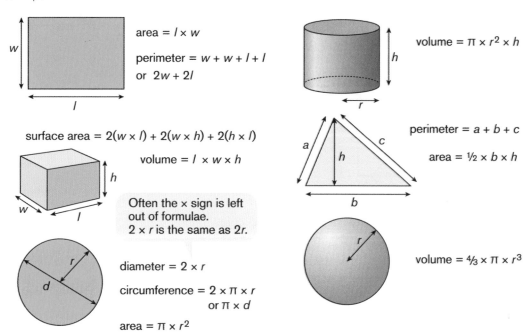

area = $l \times w$

perimeter = $w + w + l + l$
or $2w + 2l$

volume = $\pi \times r^2 \times h$

surface area = $2(w \times l) + 2(w \times h) + 2(h \times l)$

volume = $l \times w \times h$

perimeter = $a + b + c$

area = $\frac{1}{2} \times b \times h$

Often the × sign is left out of formulae.
$2 \times r$ is the same as $2r$.

diameter = $2 \times r$

circumference = $2 \times \pi \times r$
or $\pi \times d$

area = $\pi \times r^2$

volume = $\frac{4}{3} \times \pi \times r^3$

Figure A: Calculating perimeters, areas and volumes.

Scientists often calculate the **surface area : volume ratio (SA:V)** of an object. This is an object's surface area divided by its volume. Despite its name, it is written as a single integer or decimal number and not in ratio form (see S3). For example, the SA:V ratio of a cube would be stated as being 6 and not 6:1.

Shapes that involve circles use a number called **pi** (written as π). This number never changes (it is a **mathematical constant**) and is the **circumference** of any circle divided by its **diameter**. Another way of saying this is that any circle's circumference is π times its diameter. Or, the ratio of any circle's circumference to its diameter is π:1.

Apply your skills ▶▶ **B1, B8, B10** ▶▶ **C3, C6, C8** ▶▶ **P3**

DEVELOPMENT OF SCIENTIFIC THEORIES

The **scientific method** is a process (series of steps) that scientists use to show whether their ideas are correct or not. A man called Ibn al Haytham (born in Basra, which is now in Iraq, in 965) laid the foundations of this method. He is often called 'the first scientist' because he was the first thinker to test ideas using scientific experiments. Figure B is a **flow chart** showing the steps in the scientific method.

Figure A: Ibn al Haytham (965–1039).

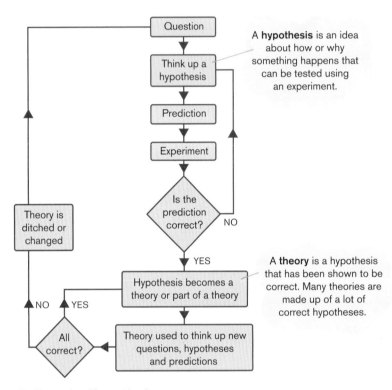

A **hypothesis** is an idea about how or why something happens that can be tested using an experiment.

A **theory** is a hypothesis that has been shown to be correct. Many theories are made up of a lot of correct hypotheses.

Figure B: The scientific method.

There are variations to the scientific method. For example, astronomers must test their predictions using observations (you can't perform experiments with planets and stars!).

The scientific method is often used to produce and to test **models**. A model is anything that represents a thing or a process in a way that helps us to understand it. There is more on models in S31.

Theories and models change over time as further experiments and observations produce new data. For example, Isaac Newton proposed a set of mathematical equations that were used to explain the motion of the planets. His model worked for all the planets apart from Mercury. Astronomers accounted for this by hypothesising that there must be another planet near Mercury's orbit (which they called Vulcan). This planet, however, could not be found. In the 20th century, Einstein came up with some new equations to explain the motion of the planets, which correctly modelled the motion of Mercury.

Apply your skills ▸▸ **B3** ▸▸ **C9** ▸▸ **P1, P2**

S34 HYPOTHESES AND THEORIES

A **hypothesis** is an idea about how or why something happens, which can be tested by experiment or observation. A hypothesis that has been tested and shown to be correct (or not to be wrong) becomes a **theory**. So, a theory is an idea about how or why certain things happen, with **evidence** to support it.

Often a theory is produced from many hypotheses, that all have supporting evidence. For example, the **kinetic theory** has many hypotheses that deal with the different states of matter (solids, liquids, gases) when they are heated or squashed or hit.

A theory:

* allows predictions to be made
* explains all the observations
* may explain other observations that weren't thought to be linked to the theory
* can be tested.

Apply your skills ⏩ B3 ⏩ C2, C6 ⏩ P6

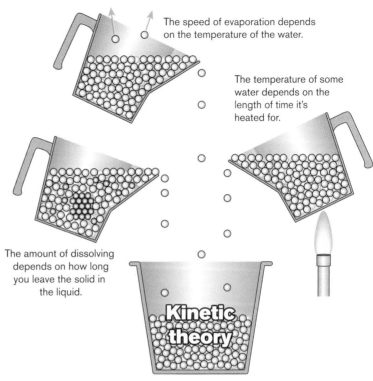

The speed of evaporation depends on the temperature of the water.

The temperature of some water depends on the length of time it's heated for.

The amount of dissolving depends on how long you leave the solid in the liquid.

Kinetic theory

Figure A: The kinetic theory contains many hypotheses.

S35 SCIENTIFIC QUESTIONS

Scientific questions are questions that can be answered using information from experiments. Questions like 'Why do plants grow towards light?' or 'How do rainbows form?' are scientific questions.

Not all scientific questions have answers because we don't have enough good information. This may be because not enough experiments have been done yet or because the experiments are too expensive or are impossible to do with current technology.

Questions involving attitudes, or **morals** or **ethics**, are not scientific questions. For example, questions such as 'Should I go to the party?' and 'Why is blue the best colour?' are not scientific because you can't do experiments to answer them.

Apply your skills ⏩ B9 ⏩ P3, P4

Hmm. What makes an apple detach from the tree when it's ripe?

Figure A: What if Newton had been a biologist?

S36 BENEFITS, DRAWBACKS AND RISKS

A **benefit** is a good thing that comes from something. The opposite of a benefit is a **drawback**. A **risk** is how likely it is that a drawback will cause harm.

Scientific research allows the development of new technologies and ways of doing things. These all have benefits, drawbacks and risks.

Apply your skills ▶▶ **B9** ▶▶ **C3, C5, C8** ▶▶ **P1, P3**

Benefits	Drawbacks	Risks
• discovery of what the Moon is like • study the effects of microgravity on humans • invention of new materials and technologies (e.g. memory foam)	• expense • danger • harm to the environment	• astronauts quite likely to die • animals likely to be killed by launch • ozone layer may be damaged by fuels

A few of the benefits, drawbacks and risks of sending people to the Moon.

S37 DECISIONS: RISKS

Everything we do carries a **risk** of harm. The size of a risk is sometimes measured by taking a large **sample** and counting the number of times something causes harm.

New technologies can create new **hazards** or drawbacks with risks of harm. For example, older women may use hormone replacement therapy (HRT), which increases the risk of breast cancer. When people make decisions (for example, to take HRT) they consider the benefits, the level of risk and the consequences if the worst happens.

Risk of breast cancer in women aged 50–70	per 1000	percentage
… without HRT	45	4.5%
… with HRT	47	4.7%

When people make decisions for themselves, they accept a certain level of risk. They are also more likely to decide to do things if the possible harm won't last very long (for example, the risk of a broken arm) compared to when the possible harm is long-lasting.

Public bodies (such as councils) compare the benefits and drawbacks of a change, and assess the levels of risks they think people should accept (for example, the risk to health of building an incinerator). However, people are less likely to accept levels of risk when they have not been given a choice.

People often think that the level of risk is higher than it actually is, particularly when the risk is from things that are unfamiliar, invisible, or may cause long-term harm.

Apply your skills ▶▶ **B2** ▶▶ **C7** ▶▶ **P7, P8**

To decide whether to use a new scientific development, people must consider all the things that might happen, both good and bad, including:

* the financial costs
* the effect on the environment
* the effect on people (both individuals and different groups)
* ethics (see below).

Figure A shows some things to consider when planning a new quarry. The colours of the bullets relate to the groupings in in the list above.

Figure A: Some factors to consider when deciding whether to build a new quarry. Some factors are more important than others.

Not all of the factors used to make a decision can be based on science – there are limitations to how much science can be used.

A decision is usually explained in a report, which:

* considers if the benefits are greater than the drawbacks
* considers if any risks can be reduced
* is based on **valid**, **repeatable** and **reproducible** evidence (see S49), not on the attitudes of a few people, or on myths or rumours.

If a decision is likely to be unpopular, evidence against the decision might be purposely hidden or the importance of favourable evidence might be over-stated.

Decisions can take time, with long reports being written. Sometimes a **public enquiry** is held, in which everyone can give their point of view. In the end, not everybody will agree with a decision, especially those at risk from harm or those who don't benefit.

Figure B: Scientists present a decision at a press conference.

Some new technologies cause ethical and moral problems. **Morals** involve actions that a person thinks is right or wrong. **Ethics** involve actions that a group of people agree are right or wrong (for example, stealing is wrong). People may disagree over ethical issues (for example, experiments on human embryos). Often an ethical decision is based on whatever leads to the best outcome for the greatest number of people.

Ethics and risks mean that many areas of science are governed by rules. These are decided upon by experts and are often made into laws.

Apply your skills ▶▶ **B4, B6, B7** ▶▶ **C4, C8** ▶▶ **P1, P4, P5**

Scientists tell others about their ideas by giving talks at conferences and writing **papers**. A paper has the same form as an investigation report:

* abstract – an overview
* introduction – discusses other research in this area
* methods – what was done and how it was done
* results
* conclusion – what the results show and why
* references – all the papers, websites etc. used to write the paper.

A paper is sent to a **journal** (a magazine for scientific papers).

The editors of journals send papers that look good to experts in the same subject as the paper. These scientists **evaluate** the investigations and check that the **conclusions** can be drawn from the results. They then say whether the paper should be published or if changes are needed. This is called **peer review**.

A scientist may not agree with the conclusions in a paper since scientists often develop different ideas to explain the same data. Only further experiments can settle the matter but the paper can still be published. However, the reviewing scientist should check that the conclusions are **valid** (they can be drawn from the results).

A reviewing scientist also checks that the method is valid (and tests what it is supposed to be testing) and that the results are not **biased** (shifted in a certain direction). Poor investigation design or poor measuring can cause bias. However, sometimes scientists cause bias on purpose to:

* please people who employ them
* become famous
* make money.

A theory is not discarded until a new theory that fully explains all the observations has been developed. This involves many investigations and papers – not just one (which may be biased or contain **anomalous results**). Scientists tend not to believe evidence that is not **repeatable** or **reproducible**. However, evidence may be given more importance if it's from a famous or well-respected scientist.

Apply your skills ▶▶ **B3, B7, B9** ▶▶ **C1, C2, C9** ▶▶ **P1, P6**

Figure A: The journal *Science* is published every week.

scientist sends paper to journal

checked by editor and sent to other scientists for comments

based on feedback from the scientists, the editor decides whether to publish

Figure B: The peer review process.

To answer a scientific question a scientist will think up an idea about how or why something happens. This is a **hypothesis**. Scientists must think creatively since a hypothesis needs to explain the observations that led to the original question.

The phrase 'depends on' is often used when writing a hypothesis. For example, we could hypothesise that the height of a plant *depends on* the amount of fertiliser it is given.

The hypothesis can be used to write a **prediction** about what will happen in a certain experiment when the hypothesis is tested. For example, *if* plant A is given twice as much fertiliser as plant B *then* plant A will grow taller.

A scientist will give reasons for thinking that a hypothesis and prediction are correct.

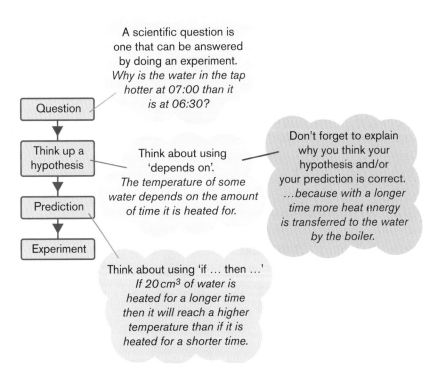

Figure A: The steps taken when testing an idea.

WEA Notice that:

✻ *'depends on'* can be useful when writing a hypothesis
✻ *'if ... then ...'* can be useful when writing a prediction.

Apply your skills ►► B2, B3, B4, B9 ►► C2, C6, C10 ►► P1, P3, P6

S41 QUALITATIVE AND QUANTITATIVE DATA

Evidence from experimentation or observation is usually in the form of **data** (numbers or words that can be organised to give information). When data is given as numbers it comes with something telling you what the numbers mean (for example, a unit of measurement, or percentage). Numbers with meanings are **values**. '5 cm' is a value but '5' is not because you don't know what the 5 refers to.

Data in number form is **quantitative data** and there are two forms: **continuous** and **discrete**.

- **Continuous data** is where each value can be any number between two limits.
- **Discrete data** is where a value can only be one of a limited choice of numbers.

Data not in number form is **qualitative data** or **categoric data** (data put into categories). **Name Check!**

Don't confuse quantitative and qualitative. Numbers come in 'quantities' and so they are quantitative data.

Apply your skills ▶▶ **B2, B5** ▶▶ **C5** ▶▶ **P2**

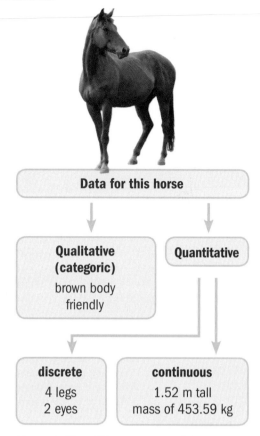

Data for this horse

→ Qualitative (categoric)
brown body
friendly

→ Quantitative

→ discrete
4 legs
2 eyes

→ continuous
1.52 m tall
mass of 453.59 kg

Figure A: The different types of data.

S42 PRIMARY AND SECONDARY DATA

- **Primary data** (or **primary evidence**) **Name Check!** is data that you collect yourself, by doing investigations.
- **Secondary data** (or **secondary evidence**) **Name Check!** is data collected by others.

Type of data	Advantages
Primary	You control the quality of the data
	You control what data is collected
	The data is up to date
	The data is relevant to your needs
Secondary	Quicker to obtain
	Easier to obtain
	Doesn't cost much to get it

Apply your skills ▶▶ **B3, B5** ▶▶ **C2** ▶▶ **P5**

S43 VARIABLES AND FAIR TESTS

A **variable** is a factor that can change. In an investigation, the variable that is changed is the **input variable** or **independent variable**. **Name Check!** The variable you measure is the **output variable** or **dependent variable**. **Name Check!** The dependent variable *depends* on the independent variable.

In a hypothesis, if you can't work out which variable is which, try them both ways round using the word 'depends'. It only makes sense when the dependent variable comes first. For example:

* the length of time water is heated depends on its temperature rise ✗
* the temperature rise depends on the length of time the water is heated. ✔

A **control variable** is a variable that could change during an investigation and affect the dependent variable. So you try to stop control variables changing. A **fair test** is when all these variables are successfully controlled.

Apply your skills ▶ B2, B4, B10 ▶ C1, C3, C6, C8, C10 ▶ P4, P7, P9

Hypothesis: The current depends on the number of cells. Prediction: If the number of cells is increased then the current will increase.

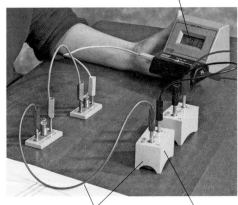

dependent variable: the current

control variables:
i) same type of cell used
ii) rest of circuit kept the same

independent variable: the number of cells

Figure A: Different sorts of variables.

S44 CHOOSING MATERIALS AND METHODS

The table lists some standard pieces of equipment used to measure certain quantities or to carry out certain processes in science. Make sure you know what they all are and how to use them.

Quantity or process	Apparatus	Comments
angle	protractor	
current	ammeter	Connected in series with a component in a circuit.
filtering	sieve filter paper	Filter paper (in a filter funnel) is used to separate finer particles than a sieve can separate, but takes longer.
force	forcemeter	Also called a newtonmeter because its scale is marked in newtons.
heating	Bunsen burner water bath	A water bath may be safer to use than a Bunsen burner and will allow much greater control over temperature. But a water bath will not reach the high temperatures needed for some reactions.
length	tape measure ruler calipers	Calipers are only used for very small distances where it is important to measure to a greater **resolution** than ±1 mm.

Continued overleaf →

Quantity or process	Apparatus	Comments
mass	bathroom scales spring balance electronic balance	Bathroom scales are only suitable for measuring large masses. Electronic balances have a high resolution (to 1, 2 or 3 decimal places).
pH	chemical indicator pH probe	Different indicators cover different ranges of pH. You need to choose one with an appropriate range for your experiment. A pH probe is more difficult to set up but has a higher resolution than an indicator.
sampling organisms	sweep net quadrat pitfall trap	The choice of sampling method will depend on the type of organisms to be studied.
temperature	thermometer temperature probe	A temperature probe is more difficult to set up and use but has a higher resolution than a thermometer. Different thermometers have different measurement ranges (e.g. 0–100 °C).
time	clock stopwatch/stopclock light gates and timer	A stopwatch or stopclock has a high resolution but is prone to human reaction time which causes random errors in the measured time due to starting or stopping too early or too late. Light gates can be used to start and stop timers automatically.
voltage	voltmeter	Connected in parallel ('across') a component in a circuit.
volume	measuring cylinder syringe burette gas syringe displacement can	The resolution of a burette, measuring cylinder or gas syringe depends on the smallest scale division of the instrument, for example ± 1 cm^3 or ± 0.1 cm^3. The range of the measuring cylinder or burette depends on its capacity (the maximum volume that can be measured). A displacement can is used to measure the volume of an irregularly shaped solid.

There is often a choice of apparatus to be made. You need to decide on the most appropriate choice of apparatus for your experiment, based on:

- how accurate you need the measurements to be
- the expected **range** of measurements
- how much time you have
- ease of use
- equipment availability.

You also need to think about how best to record measurements made with different pieces of apparatus. For example, when using a temperature probe you could use a datalogger or design a table (see S7) in which to record temperatures taken at regular intervals.

Apply your skills ⏩ B8 ⏩ C6

S45 TRIAL RUNS

When planning an investigation it is often difficult to know:
* the range of measurements to make (highest and lowest)
* what interval to have between the measurements
* how many measurements to make.

In a **trial run** you carry out an investigation quickly, taking a few measurements over a large range. It lets you see a rough pattern, so you can work out the range and number of measurements you need to be sure of this pattern in the real investigation.

A trial run also means that you won't waste time taking unnecessary measurements in your investigation (for example, by having too small an interval between measurements). It lets you make sure that your method works and that you have the correct apparatus (including measuring devices of the right accuracy and **resolution**; see S51). It also allows you to check that your investigation is safe.

Apply your skills ▶▶ B1, B4 ▶▶ C8 ▶▶ P3, P9

Figure A: Some trial runs use scale models to test ideas before building a full-sized machine to test. This model aeroplane is being tested in a vertical wind tunnel.

S46 CONTROLS

If a fair test is not possible (if there are too many control variables) a **control experiment** can be done, which has a **control** or **control group**. **Name Check!** A control is treated in the same way as the rest of the investigation but without the independent variable.

The control for the experiment in S43 would be a circuit without a cell (although we don't use controls if a fair test is possible).

Controls are used when investigating living things because they vary so much. The organisms are divided into groups, making sure that the groups are as similar to each other as possible. The independent variable is changed for the groups, but one group is a control. The results from the control make it easier to see if changes are due only to the independent variable.

Apply your skills ▶▶ B2, B6, B9 ▶▶ P1, P4

Hypothesis: The time it takes for a headache to go depends on the amount of drug X taken.
Prediction: If more doses of drug X are taken then people will feel better more quickly.

Group A: 1 dose of drug X

Group A: 2 doses of drug X

Group A: 3 doses of drug X

Control: no drug X

Figure A: Controls are used in drug tests.

S47 SAFETY: RISKS AND HAZARDS

You must plan safe investigations, which means thinking about the **hazards** of the apparatus, chemicals and methods that you want to use. A hazard is when something can cause a certain type of harm.

A **risk** is the chance of harm occurring from a hazard. You need to plan to reduce the risks from hazards. The table gives some examples.

Hazard	Example of action to reduce risk
broken glass	reporting it to the teacher to get it cleaned up
chemicals	not breathing in dust
heating things	wearing eye protection
using electricity	switching off a power pack before altering a circuit
spills	mopping up immediately
living things	using disinfectants to kill micro-organisms

explosive · harmful to health · flammable · very toxic (poisonous) · harmful to breathing system · corrosive (attacks skin)

Figure A: Symbols are used to show the hazards of things, especially chemicals.

Apply your skills ▶ B8 ▶▶ C4, C8, C10 ▶▶ P3, P4, P7

S48 ANOMALOUS RESULTS AND OUTLIERS

An **anomalous result** or outlier **Name Check!** is a measurement that does not fit into the pattern of the other results. Anomalous results should be examined to work out how they may have been caused, since they may be caused by an error.

If you can explain why an anomalous result may be wrong, ignore it when working on your data. If you can't explain it then repeat it, if you can, but leave it in your results.

Apply your skills ▶ B2, B6 ▶▶ C3, C8, C10 ▶▶ P4, P5

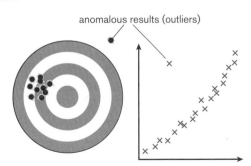

anomalous results (outliers)

Figure A: Spotting anomalous results.

S49 REPEATABILITY AND REPRODUCIBILITY

You can be more sure of conclusions from an investigation if you have more data.

You should check measurements by repeating them. Measurements that are more or less the same (**precise**) when you repeat them are called **repeatable**. Repeatable measurements allow you to be more confident that you have good quality data.

Measurements that are very similar when repeated by others are **reproducible**. Reproducible measurements allow you to be even more confident of your data.

Good quality data:

- ✕ have repeatable (and preferably reproducible) measurements
- ✕ have many measurements over a **range** that is large enough to see a pattern
- ✕ have few **anomalous results**.

Good quality data that is repeatable and reproducible allow you to draw a firm **conclusion**.

Apply your skills ▶▶ **B2, B7** ▶▶ **C1, C6** ▶▶ **P1, P2**

Figure A: Using the same make of bow, two archers both had repeatable shots. However, the two archers did not repeat each others' shots (the shots were not reproducible).

S50 ACCURACY, PRECISION AND UNCERTAINTY

Accuracy describes how close a measurement (or the average of a set of measurements) is to the true value.

Precision is how well grouped together a set of measurements are. If they are all close together, the measurements are described as being precise.

Less precise measurements lead to greater **uncertainty** about the true value. An estimate of the level of uncertainty can be written next to a value, such as:

4.0 g ± 0.2 g

The 4.0 g is the mean value and the 0.2 g shows the upper and lower limits of the set of readings (3.8 to 4.2 g).

Apply your skills ▶▶ **B6** ▶▶ **C5, C6** ▶▶ **P7, P10**

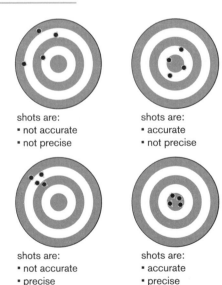

shots are:
- not accurate
- not precise

shots are:
- accurate
- not precise

shots are:
- not accurate
- precise

shots are:
- accurate
- precise

Figure A: Accuracy and precision are different.

S51 ERRORS IN MEASUREMENTS

There will always be variation in a measurement if you repeat it. There is also a limit to how small a change any device can detect. This is its **resolution** or sensitivity. **Name Check!**

Look at Figure A. Using a measuring instrument with a higher resolution we can detect that one shot has not hit the very centre of the target. We cannot detect this with the lower resolution instrument – the shots are both in its most central square.

Human error happens when people make mistakes. It happens, even when measuring is done correctly, if an investigation is not carried out in the right way.

Random errors are when there is no pattern to the errors. It happens when repeated parts of an investigation are not done in exactly the same way. Measurements with random errors are usually not grouped together – they are not **precise**.

Systematic errors are caused when repeated parts of an investigation are done in the same way but that way is not correct. Measurements with systematic errors are often grouped but are not close to the real value – they are precise but not **accurate**.

Apply your skills ▶▶ **B5** ▶▶ **C1, C10** ▶▶ **P2, P3, P6, P9, P10**

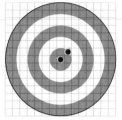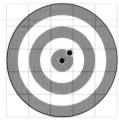

higher resolution – the instrument can detect the difference in position between the two shots

lower resolution – the instrument cannot detect the difference between the two shots

Figure A: Resolution.

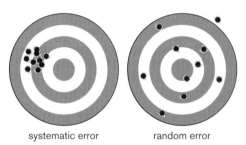

systematic error   random error

Figure B: Random and systematic errors.

S52 CONCLUSIONS

A **conclusion** is a decision that is made after considering all the evidence. To reach a conclusion you may need to present your data in a chart or graph (see S23), you may need to process your data using equations or interconversions (see S19, S20, S21), and you may need to identify patterns (see S16).

A conclusion section in an investigation report must be organised to make it easy to follow. For example:
* state what you believe you have found out
* explain how your results support this
* say whether your concluding statement agrees with your **prediction** or how they differ.

Your conclusion must also:
* be **valid**. It must be drawn from the results, which must also be valid (see S53). It must not say things that cannot be worked out from the results.
* use scientific words (like the ones in bold on these pages).

Apply your skills ▶▶ **B5, B6, B7, B10** ▶▶ **C3, C10** ▶▶ **P1, P5**

S53 VALIDITY

Something is **valid** if it does what it's meant to do. An investigation is valid if the results let you answer the original question. A **fair test** is valid because you only measure the effects of the independent variable. You keep the control variables the same so they don't affect the dependent variable (see S43).

Figure A: A fencer intends to push the foil onto the darker grey part of the opponent's jacket. A hit is only valid if this happens.

Results are valid if the measurements are what was meant to be measured and can be repeated. If control variables are not kept the same, the measurements are not valid because the dependent variable is affected by the independent variable *and* by other variables.

If the outcome of an investigation is unexpected, you need to check that the method is valid and the results are valid.

Apply your skills ⏩ **B4, B8, B9** ⏩ **C3** ⏩ **P5**

S54 EVALUATING

An **evaluation** is a look at how well something does its job. When you evaluate your investigations, or those of others, ask yourself questions about the method, results and conclusions.

Evaluate the method and the results:
- Is the investigation **valid**? (Do the results let you answer the question?)
- Are the measurements valid? (Do they measure what they were intended to?)
- How **precise** are the results? How **repeatable** are they? (Were there lots of **anomalous results**?)
- How **reproducible** are the results?
- Why did any anomalous results appear? (Think about possible sources of **random error**.)

Evaluate the conclusion:
- Were the results **accurate** enough to draw a conclusion that you can be sure of? (Think about possible sources of **systematic error**.)
- Were there enough results to draw a conclusion that you can be sure of?
- Are the results and/or conclusions unbiased? (Or is there **bias** – an attempt to persuade people that the results show something they don't?)
- Is the conclusion valid? (Has it been drawn from only the results?)
- Have any false **assumptions** been made? (Assumptions are things that everyone accepts as true so you don't test them, for example the Earth is like a ball.)

You should try to **justify** your answers.

Apply your skills ⏩ **B5, B9** ⏩ **C3, C10** ⏩ **P7, P9**

It's useful if everyone uses the same system for measurements. All scientists use the 'Système International' or **SI system**. Six basic units in this system are:

Quantity	Unit name	Symbol
length	metre	m
mass	gram	g
time	second	s
temperature	degree Celsius or kelvin	°C or K
current	ampere, amp	A
amount of substance	mole	mol

Additions are put at the front of the units to make them bigger or smaller:

Name addition	Symbol addition	Meaning (words)	Meaning (numbers)	Meaning (standard form)
nano-	n	one thousand millionth	0.000000001	$\times 10^{-9}$
micro-	µ	one millionth	0.000001	$\times 10^{-6}$
milli-	m	one thousandth	0.001	$\times 10^{-3}$
centi-	c	one hundredth	0.01	$\times 10^{-2}$
deci-	d	one tenth	0.1	$\times 10^{-1}$
kilo-	k	× one thousand	1000	$\times 10^{3}$
mega-	M	× one million	1 000 000	$\times 10^{6}$
giga-	G	× one thousand million	1 000 000 000	$\times 10^{9}$
tera-	T	× one million million	1 000 000 000 000	$\times 10^{12}$

Other units are formed from the basic units in the first table above:

Quantity	Example units
area	m^2
volume	m^3
density	g/m^3
speed, velocity	m/s
	km/h
acceleration	m/s^2
momentum	kg m/s
force	N (newton)
pressure	N/m^2
	$1\,N/m^2 = 1\,Pa$ (pascal)

Quantity	Example units
power	W (watt)
energy	J (joule)
	Wh (watt-hour)
frequency	Hz (hertz)
power of a lens	dioptre
potential difference	V (volt)
resistance	Ω (ohm)
astronomical distance	ly (light-year)
gravitational field strength	N/kg
magnetic flux density	T (tesla)

Apply your skills ▶▶ **B5, B10** ▶▶ **C2, C3, C4, C8, C9** ▶▶ **P2, P5, P8**

S56 COMPOUND UNITS

Compound units or **compound measures** **Name Check!** are formed from two or more other units. For example, the unit for work done is the newton metre, N m.

A common unit for speed uses 'metres' and 'seconds'. These units are shown together in a form that means 'the number of metres travelled in one second' or 'metres per second'. We show 'per' as a slash '/'. The unit is m/s.

HIGHER Compound units can be written with a negative **index** (because it makes some calculations easier to do). A negative index means 1 divided by the positive power of the number. So, x^{-1} is the same as $1/x$, x^{-2} is $1/x^2$ and x^{-3} is $1/x^3$. For units, we can therefore write m/s as m s^{-1} and g/cm^3 will become g cm^{-3}.

Apply your skills ⏩ **B6, B10** ⏩ **C3, C5, C6, C9** ⏩ **P2, P8**

S57 INDEX FORM

Some SI units are formed from multiplying one unit a number of times. For example, the **area** of a rectangle is worked out by multiplying its length by its width. If both measurements are in metres the unit for area will be 'square metres', which just means 'metres multiplied by metres'. We show the number of 'metres' that have been multiplied together using a **power** or **index** **Name Check!** which is a number written after and above the unit symbol: m^2. Figure A shows another example.

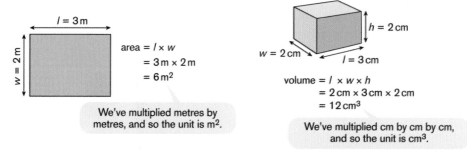

area = $l \times w$
    = 3 m × 2 m
    = 6 m^2

We've multiplied metres by metres, and so the unit is m^2.

volume = $l \times w \times h$
     = 2 cm × 3 cm × 2 cm
     = 12 cm^3

We've multiplied cm by cm by cm, and so the unit is cm^3.

Figure A: Using index form saves time and is recognised by all scientists.

Numbers can also have indices. For example, 5^2 means 5×5. This is said as 'five squared'. 5^3 means $5 \times 5 \times 5$ and is said as 'five cubed'.

Apply your skills ⏩ **B1, B5, B10** ⏩ **C2** ⏩ **P2, P3**

Words in a question that tell you how to answer the question are called command words. It's important that you know what they all mean. The table will help you.

Command	Notes
Assess **Name Check!**	See 'Evaluate' below.
Calculate	Work out an answer using numbers. Always show your working. Always put in the units.
Compare	Describe the differences or similarities between things, or their advantages or drawbacks. You need to point out the similarities/differences between or advantages/drawbacks of *all* the things.
Compare and contrast	Describe the similarities *and* differences between the things given in the question.
Complete	Fill in answers in a space or finish writing a sentence.
Define	State briefly what something means.
Describe	Recall facts in an accurate way or say what a diagram or graph shows (e.g. what trend you can see).
Determine	Use information given in the question to work out a numerical answer.
Discuss	Identify the different points made about a certain issue or problem. Then build up your own argument about the issue.
Estimate	Make a rough calculation.
Evaluate **Name Check!**	Say how good or poor something is based on looking at a series of points (criteria). If you are asked to evaluate more than one thing then you need to compare the things (see 'Compare' above) and then state which of the things is best, with reasons why you think that.
Explain	State the reasons why something happens. You must explain the links between your reasons and what you are explaining – don't just write down a list of reasons.
Give	See 'Write down' below.
Identify	Look at some data or text and pick out a certain part.
Illustrate	Give examples in an explanation or description. A good way to do this is to explain/describe something and then use the words 'For example' to introduce an example.
Justify	Give some evidence or reasoning to support an answer.
List	Write down key points in a brief way.
Name	See 'Write down' below.
Outline	State the main points of an argument or of how something happens. This can be done a list of bullet points.

Command	Notes
Show that	Demonstrate that a statement in the question is correct.
State	See 'Write down' below.
Suggest	Use your scientific knowledge to work out what is happening in an unfamiliar situation.
Summarise	See 'Outline' above.
Use the information	Use the information given to answer the question. You won't get any marks unless you use the information given.
Write down	State a fact or give an example. If a question asks for two examples (or has two marks) write down only two. Otherwise, you risk writing something that is incorrect and losing marks.

Apply your skills ➠ **B1–B10** ➠ **C1–C10** ➠ **P1–P10**

S59 LONG-ANSWER QUESTIONS

Some written answers to questions in an exam will be worth 4, 5 or 6 marks. You need to plan your answer to these before writing.

When you look at an exam question, try to pick out its key features. Here are some ideas to consider when you first look at a question:

- Make sure you understand the command word (see S58).
- Circle any key terms.
- Underline any information or figures that the question wants you to use.
- Look carefully at any diagrams, charts or graphs and ask yourself 'What is this telling me?'.

WEA Planning a long answer:

- Make some notes about information and key words you want to include. You could draw lines between words and phrases to link them or arrange them into a table.
- Number your notes to show the order in which you will cover the ideas, or write them out into a list or a flow chart.
- Cross through the notes you have made as you write your answer.
- Think about how you will *use* any information given to you. Do not simply write out information that is already in the question – that does not score any marks.

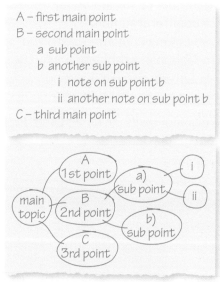

```
A – first main point
B – second main point
    a  sub point
    b  another sub point
        i  note on sub point b
        ii another note on sub point b
C – third main point
```

Figure A: There are many different ways of structuring your notes.

Continued overleaf →

- Check that you have answered all parts of the question.
- Check that you have used the examples or information that you were asked to use.
- Check your spelling, punctuation and grammar.
- Check that you have crossed out any notes that you used to help plan your answer.

Information on how long-answer questions are marked can be found on pages 86–89.

Apply your skills ▶▶ **B1, B2, B3, B4, B5, B6, B7, B9, B10**
▶▶ **C1, C2, C3, C4, C5, C7, C8, C10** ▶▶ **P1, P3, P4, P5, P6, P7, P8, P9**

S60 DISCUSSION QUESTIONS

Some longer-answer questions ask you to **discuss** an issue or a problem. You are expected to:
- identify the issues or points of view
- investigate the issue by saying what you think about the various points.

A good way of investigating the issue is often to structure your answer as an **argument**.

An argument is a way of telling people what you think and why you think it. A good argument will have the following structure:
- a statement of what you think (your point of view)
- an explanation of why you think this, with supporting evidence (making use of the points you have identified about the issue)
- a **counterargument** – an explanation of why others may not agree with you
- a response – explaining why the counterargument is wrong
- a summary – stating your point of view again.

Apply your skills ▶▶ **B5** ▶▶ **C4** ▶▶ **P3, P4, P7**

Figure A: An argument.

All rhinos have hair but the Sumatran rhino is very hairy.

Due to loss of habitat and poaching, the numbers of all rhinos have decreased. There may be fewer than 100 Sumatran rhinos left, living in a small area. There are about 14 500 white rhinos spread across huge open areas in Africa.

To see if conservation efforts are working, scientists count the rhinos each year. For small numbers of animals in small areas, they try to count each one. Automatic cameras are used to photograph Sumatran rhinos. For white rhinos, scientists use helicopters to count the numbers in one area and estimate the total.

Figure A: Sumatran rhinos live deep in Asian jungles.

Figure B: Helicopters are used to find white rhinos.

QUESTIONS

1 What is an estimate? ▶ **S12**

2 a A scientist is estimating the number of white rhinos in a park. She selects a sample area 4 km long and 5 km wide. How big is the sample area? ▶ **S32**

  b There are 51 rhinos in the sample. Estimate the number in the whole park, which is a 140 km². ▶ **S5, S12**

3 Assuming there are 100 Sumatran rhinos, what is the ratio of the number of white rhinos to Sumatran rhinos? ▶ **S3**

4 Five Sumatran rhinos are found evenly spread over 20 km². What is the probability of a rhino being found in any 1 km² in this area? ▶ **S13**

5 There are also about 60 Javan rhinos left, about 2500 black rhinos and 2650 Indian rhinos. Draw a pie chart to compare the numbers of all five species of rhino. ▶ **S11**

6 What needs to be done in a trial run before counting Sumatran rhinos? ▶ **S45**

7 How could good data be collected to see if conservation efforts for white rhinos are working? Justify your answer. ▶ **WEA** **S12, S58, S59**

Long-Answer Grade Booster

★★★ justifies answer by explaining why the methods chosen were better than alternative methods

★★★ explains why the data was collected/analysed in this way

★★★ describes how data has been collected/analysed

An advert for a drinks company showed a peanut and said: 'The Peanut (helping the sale of soft drinks since 1830)'.

When you eat salty things, like peanuts, your blood becomes a bit more salty. This can cause problems for your body. If your brain detects too much salt in your blood it causes you to become thirsty. The extra water you drink dilutes the salt.

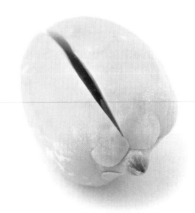

Figure A: The humble peanut.

Your cells contain dissolved substances. If cells are put in a more concentrated solution of the dissolved substances, water leaves the cells through their cell surface membranes. The cells shrivel up. The membrane lets only water molecules move through it and not the dissolved substances. The movement of water through a membrane that allows only certain molecules to pass through it is osmosis. Figure C shows a model for it.

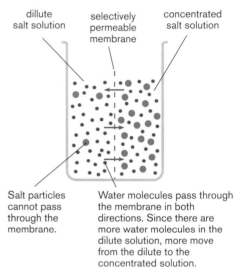

Salt particles cannot pass through the membrane.

Water molecules pass through the membrane in both directions. Since there are more water molecules in the dilute solution, more move from the dilute to the concentrated solution.

Figure C: Osmosis explained using the particle model of matter.

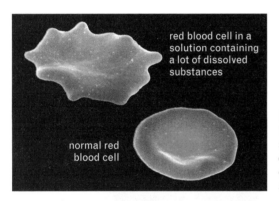

red blood cell in a solution containing a lot of dissolved substances

normal red blood cell

Figure B: Blood cells shrivel in high concentrations of dissolved substances.

QUESTIONS

1 Young children should not eat small items like peanuts because there is a risk of about 1 in 100 000 of them choking to death. Calculate this risk as a percentage.
▶▶ **S3, S37, S58**

2 Is the data in Figure C qualitative or quantitative? Explain your reasoning. ▶▶ **S41**

3 Use the information presented to define a selectively permeable membrane. ▶▶ **S58**

4 Why do scientists use models? ▶▶ **S31**

5 a An investigation is done to see what concentration of salt solution causes a human cell to start to shrivel. Explain why more than one cell is used.
▶▶ **S48, S49, S58**

b What control could be used in this investigation? ▶▶ **S46**

c Suggest two control variables for this investigation. ▶▶ **S43, S58**

6 Some blood cells are put in pure water. Explain what will happen.
▶▶ **WEA S40, S58, S59**

Long-Answer Grade Booster

★★★ uses a model to explain why the cells swell up

★★★ uses a model to describe how the cells swell up

★★★ states that the cells will change shape

Some scientists thought about how organisms might change over 5 million years if Earth gets colder. They imagined giant flightless birds called 'gannetwhales'.

In 1838 Charles Darwin (1809–1882) read an essay by Thomas Malthus (1766–1834). It said that if people had too many children some children would die. This gave Darwin the idea that organisms had more offspring than would survive. Only those best suited to the surroundings would survive and reproduce to pass their features to their offspring. If the environment changed, so would a species' features.

thick covering of dense feathers for insulation

spray half-digested fish vomit over any predators that get too close

flippers evolved from wings

Figure A: Gannetwhales might evolve from gannets (inset).

There is a lot of evidence for this idea and we are still finding more, including similar species with slightly different features living in slightly different surroundings. There is also evidence from fossils.

Figure B: Fossil leg bones provide evidence for horse evolution.

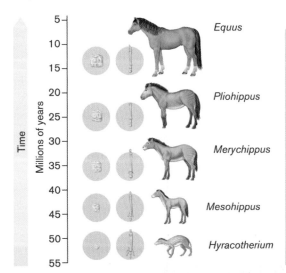

Time

Millions of years

Equus

Pliohippus

Merychippus

Mesohippus

Hyracotherium

Teeth: Grasslands replaced the forests in which horse ancestors lived. The back teeth became larger and harder, to grind tough grass.

Feet: Harder ground replaced the softer ground. The central toe became a hoof, allowing faster running on hard ground to escape predators.

Legs: More open ground replaced forest. The legs became longer allowing the animal to see further to spot predators.

QUESTIONS

1 What is the difference between a hypothesis and a theory? ▶▶ **S34, S40**

2 What gave Darwin the idea for how evolution happened? ▶▶ **S40**

3 In terms of order of magnitude, how much older is *Hyracotherium* compared with *Equus*? ▶▶ **S17**

4 a When studying evolution, scientists often use many different reports written about many fossils found by different scientists. Is this primary or secondary data? ▶▶ **S42**

   b List two disadvantages of using this type of data. ▶▶ **S42, S58**

5 How long ago did *Merychippus* exist? Give your answer in standard form. ▶▶ **S2**

6 Some fossils are very rare. What problem can this cause when using fossils to show how an organism has evolved? ▶▶ **S12**

7 How would you present data showing changes in horse leg bone length with time? Explain your choice. ▶▶ **S23, S58**

8 Explain why most scientists believe Darwin's theory of evolution. ▶▶ **WEA** **S33, S39, S58, S59**

Long-Answer Grade Booster

★★★ explains how some evidence supports the theory

★★★ describes some simple evidence that supports the idea

★★★ states that there is a lot of evidence for this theory

People who swim in the Kangal hot spring in Turkey claim that its skin-eating fish, *Garra rufa* and *Cyprinion macrostomus*, can treat skin problems.

Psoriasis (*'sore-eye-a-sis'*) is a condition in which the body sends faulty signals to skin cells, causing patches of dead skin. A pilot study tested the idea that *Garra rufa* can treat psoriasis. 67 patients had a bath with between 250 and 400 fish for 2 hours each day, as well as using a sunbed. The patients were between 10 and 75 years old.

After 3 weeks, the patients were given a 'PASI score' out of 72 (which measures how bad psoriasis is). The mean PASI score at the start was 18.9. At the end it was 5.34. The patients were also asked questions, some of which are shown in the table.

Figure A: Fish such as *Garra rufa* feed on dead skin.

Was there a reduction in ...	Answered: 'Extremely'	Answered: 'Considerably'	Answered: 'A little'	Answered: 'Not at all'
... itching?	46	15	2	0
... pain?	48	6	0	0
... skin scaliness?	60	10	1	0

QUESTIONS

1 What hypothesis was tested in the study? ▶▶ **S40**

2 Where did the idea for this hypothesis come from? ▶▶ **S40**

3 List the dependent, independent and control variables in this study. ▶▶ **S43, S58**

4 How many significant figures are the mean PASI scores given to? ▶▶ **S5**

5 A 'pilot study' is like a trial run. State two benefits of a trial run. ▶▶ **S45, S58**

6 a Convert the figures in the table into percentages. ▶▶ **S3**

 b Draw a bar chart and three pie charts to display the percentages. ▶▶ **S9, S11**

 c Which do you think is the better way of displaying these percentages? ▶▶ **S9, S11, S23**

7 The scientists who carried out this study want to publish their results in a peer-reviewed journal. Explain the main benefit of peer review. ▶▶ **S39, S58**

8 A website advertising a *Garra rufa* treatment claims the study 'shows that people with psoriasis obtain clear benefits from treatment with *Garra rufa*'. Discuss this claim. ▶▶ **WEA** **S38, S53, S58, S59**

Long-Answer Grade Booster

★★★ explains how the strengths/weaknesses of the study affect the conclusions that can be drawn

★★★ explains why some aspects of the study are strengths/weaknesses

★★★ identifies some strengths/weaknesses of the study

Organisms survive by changing their behaviour to respond to changes in their environments. Our brains are vital for this.

Scientists think that during human evolution, brain size increased to allow more complex behaviour, such as balancing on two legs and language. Evidence comes from measuring the volumes of skulls from human ancestors (Figure A).

Figure A: The scatter graph uses data from the papers of many scientists.

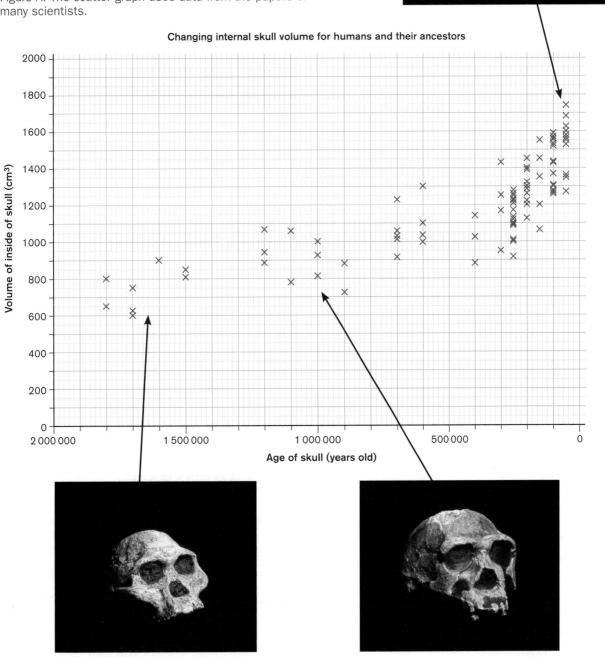

Continued overleaf →

Some scientists say that there was a steady increase in the size of the brain as modern humans evolved, which supports the idea that evolution is a slow continuous process. Other scientists say that there was a sudden increase in brain size about 200 000 years ago, which supports the idea that evolution happens in sudden jumps.

An assumption is something that you think is accepted as correct and so you don't try to show that it is correct. An assumption made by the scientists who produced the graph in Figure A is that the bigger the skull the more brain it will contain. From dissecting dead bodies, scientists have known for a long time that the inside of the skull is mainly taken up by the brain. New techniques (brain scans) have confirmed that this is true. However, brain scans have revealed some people who lead normal lives but whose brains do not fill their skulls. Brain scans like this are anomalous results.

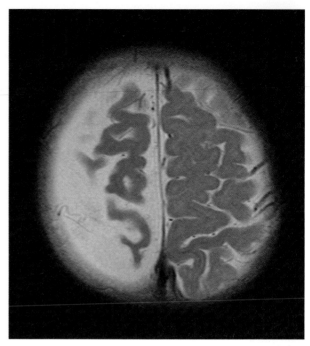

Figure B: Scan from a girl born with only half a brain. She has a normal life.

QUESTIONS

1 a What symbol is used for the unit for volume in Figure A?
 b What does the symbol mean? ▶▶ **S55, S57**

2 What are the volumes of the two oldest skulls? ▶▶ **S15**

3 a What is the range of skull volumes 600 000 years ago? ▶▶ **S6**
 b Calculate their mean. ▶▶ **S6, S58**
 c How would you show the means and the degree of uncertainty for each measurement on a graph? ▶▶ **S15**

4 Is the data in Figure A:
 a qualitative or quantitative?
 b primary or secondary? ▶▶ **S41, S42**

5 Describe the correlation in Figure A. ▶▶ **S16, S58**

6 Write '1 800 000' in standard form. ▶▶ **S2**

7 A scientist has calculated the mean volume of the skulls for each age, and now wants to calculate the rate at which skull volume increased over the years. Describe how this should be done. ▶▶ **S24, S27, S58**

8 A scientist argues that increasing brain size allowed human behaviour to become more complex. Figure A is used as evidence.
 a Suggest a counterargument.
 b Suggest two assumptions that have been made in using Figure A as evidence. ▶▶ **S54, S58, S60**

9 Why is the data used for Figure A likely to contain random errors? ▶▶ **S51**

10 Which idea in the paragraph at the top of this page do you think is more likely to be right? Justify your choice. ▶▶ **WEA** **S15, S52, S58, S59**

Long-Answer Grade Booster

★★★ refers to different possible interpretations of the data

★★★ explains choice by interpreting the data

★★★ chooses an idea and some data

NASA scientists have been studying how to grow food plants in the cramped conditions in a spacecraft carrying astronauts to Mars. They have found that LED lights don't produce much heat, so plants can be close to the lights and not wilt.

Figure A: Potato plants at NASA being grown under LED lights.

Substances in a plant take in (absorb) light and use the energy to power photosynthesis, to make food. In photosynthesis, a series of chemical reactions split water into oxygen and hydrogen. The hydrogen is then combined with carbon dioxide to make a sugar called glucose. The oxygen is released.

$$6CO_2 + 6H_2O \rightarrow C_6H_{12}O_6 + 6O_2$$

carbon dioxide + water → glucose + oxygen

Figure B: Word and symbol chemical equations for photosynthesis.

Different LEDs produce different colours and so scientists have investigated which colour is best for photosynthesis. The table shows the results from some pondweed.

Colour of light	Volume of oxygen released in 1 minute (mm^3)			
	Test 1	Test 2	Test 3	Test 4
white	66	69	67	66
blue	60	58	58	55
green	22	17	180	21
yellow	24	22	23	23
red	51	51	50	51

QUESTIONS

1 Explain why models are useful. Illustrate using an example from this page. ▶ S31, S58

2 What does the formula $C_6H_{12}O_6$ tell you about glucose? ▶ S18

3 Which figure in the table would you not use to draw conclusions? Explain why. ▶ S48

4 For which colour are the readings the most precise? ▶ S50

5 a What is the mode for the readings using blue light? ▶ S14
  b Calculate means for the data. ▶ S6, S14, S58

6 What compound unit of measurement could be used for the results? ▶ S56

7 Draw an appropriate graph or chart to display the means. ▶ S23

8 Why are controls useful? Illustrate using an example from this page. ▶ S46, S58

9 You have been asked to choose the colour of LED to be used for growing food plants on a spacecraft. Explain how you would make this choice. ▶ WEA S52, S58, S59

Long-Answer Grade Booster

★★★ explains the full range of evidence needed

★★★ interprets the evidence in the table and list some other evidence needed

★★★ chooses a colour based on the evidence in the table

In the UK, most children are vaccinated against measles.

In 1998 Dr Andrew Wakefield published a peer-reviewed paper in which he tested 12 children to find a link between a vaccination and autism (a communication disorder). Parents of 8 of the children said that their child's autism started just after an MMR vaccination (against measles, mumps and rubella).

Wakefield claimed that MMR was not safe. Some newspapers ran headlines like 'Why I wouldn't give my baby the MMR jab'.

However, one journalist discovered that solicitors working for the parents of the 12 children had paid Wakefield. The solicitors wanted to sue the makers of MMR and needed evidence against the makers.

In some later research, Wakefield needed blood from children. He obtained it by paying 22 children £5 each at his son's birthday party.

Other scientists have not been able to repeat Wakefield's findings. In 2010, he was found to have been 'dishonest' and 'unethical'. He was banned from being a doctor.

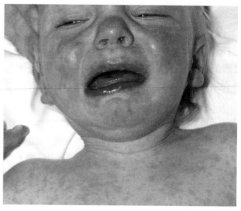

Figure A: Measles can kill.

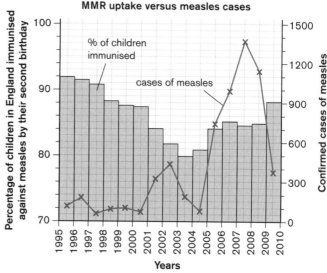

Figure B : Change in MMR uptake and change in number of measles cases, 1995–2010

QUESTIONS

1 What sort of graphs are shown above? ▶▶ **S23**

2 What is a peer-reviewed paper? ▶▶ **S39**

3 a A journalist accused Wakefield of altering his data. Suggest why Wakefield might have biased his data on purpose. ▶▶ **S12, S39, S58**

b Suggest a cause of accidental bias that might be in Wakefield's data. ▶▶ **S12, S58**

4 a Describe the correlation shown in Figure B.

b Is it is causal? Explain your reasoning. ▶▶ **S16, S58**

5 Suggest one reason why Wakefield should not have made claims about MMR based on his paper. ▶▶ **S52, S58**

6 Explain one way in which Wakefield acted unethically. ▶▶ **S38, S58**

7 How does other scientists' work go against Wakefield's conclusions? ▶▶ **S49**

8 Explain why it is important that newspapers are careful when reporting scientific claims. ▶▶ **WEA S58, S59**

Long-Answer Grade Booster

★★★ explains the effects, using an example, of influencing public opinion

★★★ describes, using an example, how a scientific claim can influence public opinion

★★★ states that a scientific claim can influence public opinion

Sampling is a way of estimating population size.

Two students measured the population of dandelions and daisies growing on a playing field. They drew a map of the field and then took samples of the area using a stiff wire frame measuring 0.5 m by 0.5 m. To take the samples, they started in the middle of one end of the field and took ten paces towards the other side. They then flipped a coin to decide whether to throw the frame to the left or to the right. Their results are shown in the table.

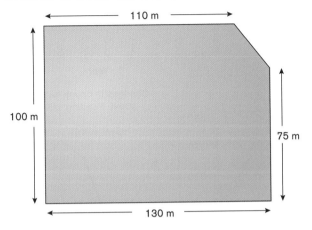

Figure A: A map of the playing field.

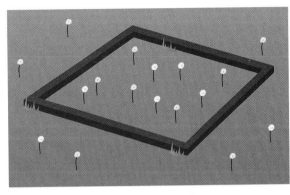

Figure B: Taking the samples.

Sample number	1	2	3	4	5	6	7	8	9	10	11	12	13	14	15
Number of dandelions	1	0	0	0	3	0	5	0	0	5	2	1	1	1	2
Number of daisies	3	5	6	7	8	0	0	7	8	4	2	2	1	1	3

QUESTIONS

1 Name the piece of apparatus used in this investigation. ▶▶ **S44**

2 a You are going to make an estimate of the populations of daisies and dandelions in the field. What is an 'estimate'? ▶▶ **S12**

b Calculate the total area of the field. ▶▶ **S32**

c Estimate the population of daisy plants in the field. ▶▶ **S12, S58**

d Estimate the population of dandelion plants in the field. ▶▶ **S12**

3 a State two factors that the students should have considered when deciding how many samples to take. ▶▶ **S12**

b Calculate the ratio of the area of the playing field to the total area that the students sampled. ▶▶ **S3**

c Do you think the students took enough samples? Explain your answer. ▶▶ **S12, S58**

4 a State one way in which the students tried to make their samples random.

b Explain why taking random samples is important.

c Suggest one way in which their method may have not been random enough.

d Describe a better way of taking random samples in an area like this. ▶▶ **S12, S58**

5 One of the students states in his conclusion for this investigation that 'dandelions are therefore more common in the sunnier parts of the field'. Explain why this conclusion is not valid. ▶▶ **S53, S58**

6 a Describe one way in which the students should have planned to stay safe when doing their investigation.

b Describe one way in which the students should have planned to avoid harming organisms in the habitat. ▶▶ **S47**

There are two main types of ultraviolet rays from the Sun that reach the surface of the Earth – UVA and UVB. Too much exposure to ultraviolet rays (especially UVB) can cause skin cancer.

Malignant melanoma is a dangerous skin cancer. There is now a 1 in 91 chance of a British man getting a malignant melanoma in a lifetime, and a 1 in 77 chance for a woman. There are increasing numbers of people getting this cancer, as shown by the graph of data for part of the first decade of this century (Figure A).

One hypothesis is that people are going on more foreign 'sunshine' holidays. Another hypothesis is that the ozone layer has been getting thinner (ozone absorbs UV rays from the Sun).

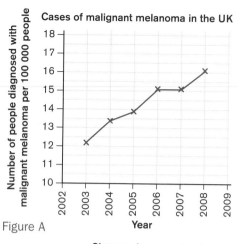

Figure A

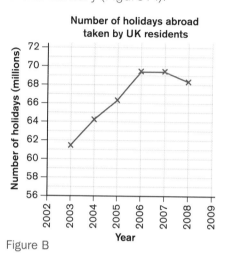

Figure B

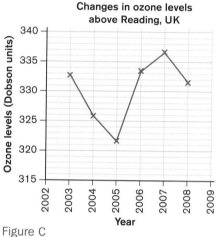

Figure C

QUESTIONS

1 What scientific question are the two hypotheses addressing? ▶ **S35**

2 a Which hypothesis does the evidence in the graphs support? Explain your choice. ▶ **S16, S40, S58**

b Suggest a problem with using this evidence to support this hypothesis. ▶ **S53, S54, S58**

3 How could you manipulate the data in the graph of ozone levels (Figure C) to make it support the hypothesis that thinning ozone is causing the increase in malignant melanomas? ▶ **S12, S24, S39**

4 a Show the number of foreign holidays taken in 2006 (from Figure B) in standard form. ▶ **S2**

b Calculate the rate of increase in the number of foreign holidays between 2003 and 2006. ▶ **S27, S58**

5 Make predictions for the cases of malignant melanoma in 2002 and 2009. ▶ **S15**

6 Use this page and Activity P7 (on page 82) to describe the benefits, drawbacks and risks of sunbathing. ▶ **S36, S58**

7 What is the percentage probability of a man getting malignant melanoma? ▶ **S13**

8 Another hypothesis is that the likelihood of you getting malignant melanoma depends on your use of sunbeds. What evidence would you collect to show this? Explain fully what control you would use for your data and why this is important. ▶ **WEA** **S10, S13, S46, S58, S59**

Long-Answer Grade Booster

★★★ explains how a control is used when processing the results

★★★ describes the control that will be used

★★★ plans a survey

Each year the National Child Measurement Programme (NCMP) measures the heights and masses of Year 6 children.

19% of Year 6 children were obese in 2015. Obese people are likely to have health problems caused by being very overweight. Causes include lack of exercise and eating lots of sugar and fat.

The BMI is an estimate of whether someone has a healthy mass:

$$\text{BMI} = \frac{\text{mass (kg)}}{\text{height}^2 \ (\text{m}^2)}$$

The table shows the different BMI categories for adults. People under 20 are still growing and so their BMIs are compared with those of people of the same age and sex (Figure B).

BMI range	Category
below 18.5	underweight
18.5 – 25	healthy
25 – 30	overweight
above 30	obese

low calorie still fruit punch flavour spring water drink with vitamins and potassium

nutrition information typical values per 100ml			
energy:	80kJ, 19kcal	fat:	0g
protein:	0g	of which saturates:	0g
carbohydrate:	4.6g	fibre:	0g
of which sugars:	4.6g	sodium:	0g

other nutrients typical values per 100ml	
vitamin C	16mg (100%*)
niacin (B3)	1.6mg (50%*)
vitamin B6	0.14mg (50%*)
vitamin B12	0.25µg (50%*)
pantothenic acid (B5)	0.6mg (50%*)
potassium	81mg (20%*)

*recommended daily amount per 500ml

this 500ml serving contains

Calories	Sugars	Fat	Saturates	Salt
95kcal	23.0g	0.0g	0.0g	0.0g
5%	26%	0%	0%	0%

% of an adult's guideline daily amount (based on a 2000 kcal diet)

Figure A: This drink contains over a quarter of the total amount of sugar recommended for a whole day.

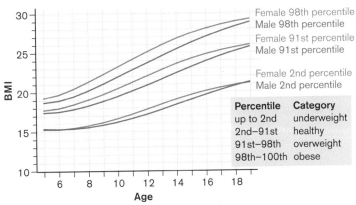

Female 98th percentile
Male 98th percentile
Female 91st percentile
Male 91st percentile
Female 2nd percentile
Male 2nd percentile

Percentile	Category
up to 2nd	underweight
2nd–91st	healthy
91st–98th	overweight
98th–100th	obese

Figure B: Percentiles used to work out BMI categories for those under 20.

QUESTIONS

1 Calculate the BMIs of the following and put each person in a category.
 a Derek is 25, has a mass of 99 kg and is 181 cm tall. ▶▶ **S21, S55**
 b **HIGHER** Jane is 12, has a mass of 43 kg and is 1.36 m tall. ▶▶ **S14**

2 a What does the symbol 'm²' mean?
  ▶▶ **S55, S57**
 b State a quantity that's often measured in m².
  ▶▶ **S32, S58**

3 The BMI is an estimate. What is an estimate?
  ▶▶ **S4**

4 Suggest one variable that is difficult to control in the NCMP. ▶▶ **S43, S58**

5 It is common for about 10% of children to opt out of being measured. Why is this a problem? ▶▶ **S12**

6 The study found a negative correlation between obesity and wealth. What does this mean? ▶▶ **S16**

7 Suggest why you think the manufacturers of the drink in Figure A first advertised it as being 'nutritious' but were then banned from using this word in the adverts.
  ▶▶ **WEA S58, S59**

Long-Answer Grade Booster

★★★ explains why the product is 'nutritious' and/or is not 'nutritious'

★★★ describes some problems with using the word 'nutritious'

★★★ states some reasons why the word 'nutritious' was used

C1 MEMORY OF WATER

A 1988 paper by Jacques Benveniste (1934–2004) claimed that water 'remembered' chemicals that had been dissolved in it. The editor of *Nature*, John Maddox, published it on condition that he went to the lab to check the work.

Benveniste took a chemical that causes certain white blood cells to lose granules that they normally contain. He mixed one part of the chemical with 9 parts of water to make a 10 × dilution (1×10^1). He repeated this to make a 100 × dilution (1×10^2) compared with the original. He made dilutions down to 1×10^{120}. The number of cells that lost their granules when added to each dilution was counted.

Benveniste's results always showed a certain pattern of peaks, meaning that some dilutions always had an effect even though they can't have contained any molecules of the chemical. When Maddox visited, the results in the graphs here were obtained. Maddox also discovered that the experiment did not always work but if this happened the results were ignored by Benveniste and his team. This was not stated in the paper.

Figure A: Are water molecules permanently affected by chemicals?

hydrogen atom

oxygen atom

water molecule

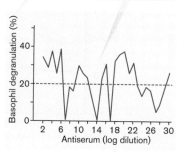

percentage of cells counted that had lost their granules

The mean values of three experiments in which the lab's cell counter knew which dilutions she was counting.

A fourth experiment in which the lab's cell counter did not know which dilution she was counting.

The number on this scale is the index (power) of ten of the dilution – e.g. 1×10^{20} is written as 20.

Figure B: Graphs from experiments during Maddox's visit.

QUESTIONS

1 a Why do we use models? ▶ **S31**
  b Suggest one good point and one poor point about the model in Figure A.
  ▶ **S31, S58**

2 What is the ratio of hydrogen to oxygen in a molecule of water? ▶ **S3**

3 Explain how a 1×10^4 dilution of the chemical would be made. ▶ **S3, S58**

4 a What is the ratio of the ranges in the two graphs in Figure B? ▶ **S3, S6**
  b What do the graphs tell you about the results in Benveniste's paper? ▶ **S49**

5 Suggest two control variables for this experiment. ▶ **S43, S58**

6 Suggest why Maddox published Benveniste's paper. ▶ **S39, S58**

7 Benveniste's original results always showed a specific pattern of peaks. Use information from this page to explain how this could have happened. ▶ **WEA** **S12, S51, S58, S59**

Long-Answer Grade Booster

★★★ explains the effects of the errors

★★★ describes the causes of the errors

★★★ states that this was due to errors

C2 THE PERIODIC TABLE

A Russian chemist, Dmitri Mendeleev, is often called the 'father of the periodic table'. But not in Germany!

Mendeleev wrote a data card for each known element, which he put in order of 'atomic weight' (the *mass* of 6.02×10^{23} atoms). He noticed a pattern and so arranged his cards as a table so that elements with similar properties lined up. He left gaps for elements that he said must exist but hadn't been found. He also swapped some elements around, saying that their 'atomic weight' measurements must be wrong. He published his table in a paper in 1869.

A German scientist, Julius Lothar Meyer, had published a similar table in 1864. It also had a gap but showed only half the known elements. In 1870, he published a complete table, working out groupings of elements from his graph shown in Figure B.

I	gap	II	III	IV	V	VI	VII		elements reversed VIII	
H 1.01										
li 6.94		Be 9.01	B 10.8	C 12.0	N 14.0	O 16.0	F 19.0			
Na 23.0		Mg 24.3	Al 27.0	Si 28.1	P 31.0	S 32.1	Cl 35.5			
K 39.1		Ca 40.1		Ti 47.9	V 50.9	Cr 52.0	Mn 54.9	Fe 55.9	Co 58.9	Ni 58.7
Cu 63.5		Zn 65.4			As 74.9	Se 79.0	Br 79.9			
Rb 85.5		Sr 88.9	Y 88.9	Zr 91.2	Nb 92.9	Mo 95.9		Ru 101	Rh 103	Pd 106
Ag 108		Cd 112	In 115	Sn 119	Sb 122	Te 128	I 127			
Ce 133		Ba 137	La 139		Ta 181	W 184		Os 194	Ir 192	Pt 195
Au 197		Hg 201	Ti 204	Pb 207	Bi 209					
				Th 232		U 238				

Figure A: An early periodic table by Dmitri Mendeleev (1834 – 1907). There's a modern version on page 126.

Lothar Meyer plotted 'atomic weight' against the volume that that mass of an element occupied. The regular peaks show a pattern of similarities.

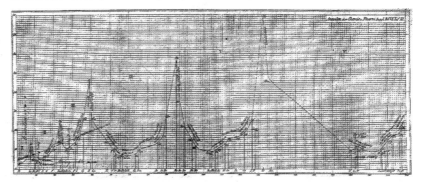

Figure B: The graph published by Julius Lothar Meyer (1830–1895).

QUESTIONS

1 State two predictions made by Mendeleev. ▶ S40, S58

2 In what way could Lothar Meyer's 1864 idea not be considered a theory? ▶ S34

3 a What element now fills the gap next to zinc (Zn) in Mendeleev's table? Use the periodic table on page 126 to help you. ▶ S7, S18

 b Why was its discovery important for Mendeleev's idea? ▶ S34

 c How did Lothar Meyer's work support Mendeleev's idea? ▶ S34

4 a What sort of graph is that in Figure B? ▶ S23

 b Suggest units for the two quantities plotted on the graph. ▶ S55, S57, S58

5 a Did Mendeleev use primary or secondary data? ▶ S42

 b Give one disadvantage of this sort of data. ▶ S42

6 State one advantage of using symbols for the elements. ▶ S18, S58

7 Write out 6.02×10^{23} as a decimal. ▶ S1, S2

8 Papers are usually peer reviewed. Describe what happens and the benefits of this process. ▶ WEA S39, S58, S59

Long-Answer Grade Booster

★★★ explains all the benefits
★★★ describes the steps and describes a benefit
★★★ describes the overall process

C3 TESTING MATERIALS

Alloys are solid mixtures of a metal with other elements. 'Alloy wheels' are pricy because they are made out of expensive metals.

Alloys have a lower density than steel and so are lighter but have the same strength. They are often used for car engines and wheels. Density is mass divided by volume.

The strengths of some alloys of titanium and manganese were investigated. One bar of each alloy was put in a machine that stretched it. The force required to make the bar yield (suddenly get thinner) was measured. The results are shown in Figure B.

Figure A: This 'alloy' is aluminium with 7.1% Si and 0.4% Mg.

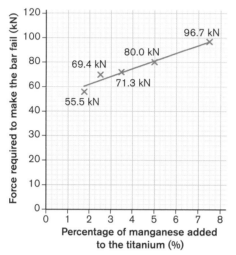

Figure B: Alloys are stronger than a pure metal when stretched.

QUESTIONS

1 a What percentage of aluminium does the wheel in Figure A contain? ▶▶ **S3**

b Draw an appropriate chart or graph to compare the percentages of the three elements. ▶▶ **S23**

2 State two ways in which you would make the testing of the alloys fair. ▶▶ **S43**

3 a Using figure B, estimate the strength (in kN) of pure titanium. ▶▶ **S4, S15, S58**

b Which results on the graph would you check? Explain your choice. ▶▶ **S48, S58**

4 State each force on the graph in newtons to 2 significant figures. ▶▶ **S5, S55**

5 a A steel block of mass 32 g measures 1 cm by 2 cm by 2 cm. Calculate its density. ▶▶ **S20, S32, S56**

b The same-sized block of aluminium alloy has a density of 2.7 g/cm³. Calculate its mass. ▶▶ **S3**

6 Suggest one benefit and one drawback of alloy wheels. ▶▶ **S36, S58**

7 Three conclusions from the experiment are:

i – The alloy with 7.5% manganese will be the best of them to use to support a bridge.

ii – The alloy with 7.5% manganese will be the best of them to use for a crane hook.

iii – We should use 20% manganese for a crane hook because it will be even stronger.

Evaluate the conclusions.
▶▶ **WEA** **S52, S53, S54, S58, S59**

Long-Answer Grade Booster

★★★ evaluates the investigation

★★★ identifies a mistake in the other conclusions

★★★ chooses a conclusion, with a reason

Fritz Haber was a German chemist who prolonged World War I.

The Germans had been making explosives using naturally occurring sodium nitrate from Chile but in the war their supply was cut off. Haber had invented a process to make a poisonous gas called ammonia, which can also be used to make explosives. Haber's method was improved to make vast amounts of ammonia for the German war effort.

Two reactions happen in the Haber process: hydrogen and nitrogen combining and ammonia splitting. Eventually these two reactions occur at the same rate, with the amount of ammonia forming being equal to the amount of ammonia splitting.

this symbol means that the reaction can go in both directions

$$N_2 + 3H_2 \rightleftharpoons 2NH_3$$

nitrogen + hydrogen ⇌ ammonia

Haber found that using high temperature, high pressure and a catalyst he could get more of the hydrogen and nitrogen to form ammonia. Haber's process is still used today and most ammonia is used to make fertilisers.

Graphs to show the percentage of ammonia produced at different temperatures and pressures

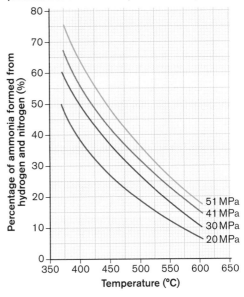

Figure A: High pressures produce more ammonia but increase costs. Lower temperatures produce more ammonia but much more slowly (which increases costs).

QUESTIONS

1 a What unit of pressure is used on the graph?
b What does this unit mean? ▶▶ **S55, S56**

2 a What is the correlation between temperature and ammonia formed?
b Is this a negative or a positive correlation? ▶▶ **S16**

3 Draw a table to show the percentages of ammonia formed by the different pressures at 500 °C. ▶▶ **S7**

4 a State one hazard of making ammonia.
b How might the risks be reduced? ▶▶ **S47**

5 a List the variables shown on the line graphs.
b Describe the variables using these words: independent, dependent, quantitative, qualitative, continuous, categoric. ▶▶ **S41, S43**

6 Today, the Haber process is carried out at 450 °C and 20 MPa. Suggest why this set of conditions has been chosen. ▶▶ **S38, S58**

7 Summarise an argument in favour of building a new ammonia factory in a town. ▶▶ **WEA** **S38, S58, S59, S60**

Long-Answer Grade Booster

★★★ uses well structured argument
★★★ explains reasons in favour and against
★★★ states a point of view with a reason

RIVER QUALITY

All rivers in England and Wales are graded from A to F.

HIGHER In order to grade each river, various factors are measured, including the concentration of ammonia (NH_3) and the percentage of dissolved oxygen. 36 water samples are collected in 3 years and the measurements are split into percentiles (see table). For example, for grade A the 10th percentile of all the dissolved oxygen readings must be at 80% (or more). The overall grade is the lowest grade given for all the measurements.

Figure A: The River Irk became a bubble bath after a soap factory spill.

Too much ammonia (NH_3) in river water is poisonous to water creatures. Ammonia levels are increased by sewage and by fertilisers. The nutrients in fertilisers also allow algae in the water to grow very quickly. When the algae die, they are broken down by bacteria, which use up the oxygen. This means that there is not enough oxygen for water creatures and so they die.

Grade	A	B	C	D	E	F
Dissolved oxygen (%) at the 10th percentile	80	70	60	50	20	<20
Ammonia (mg/dm^3)* at the 90th percentile	0.25	0.6	1.3	2.5	9.0	–

*simplified from original

QUESTIONS

1 Use the information in the table above to give one example of qualitative data and one example of quantitative data. ▶▶ **S41**

2 What does the symbol for ammonia tell you about it? ▶▶ **S18**

3 What does the symbol < mean? ▶▶ **S18**

4 Identify a compound unit of measurement used on this page. ▶▶ **S56, S58**

5 Suggest why 36 samples of a river are used to calculate a grade. ▶▶ **S12, S58**

6 State two drawbacks of using chemical fertilisers. ▶▶ **S36, S58**

7 **HIGHER** Here are the percentiles for 36 readings for a river.

Percentile	5	10	25	50	75	90	95
Dissolved oxygen (%)	75	79	84	89	92	94	94
Ammonia (mg/dm^3)	0.02	0.02	0.03	0.05	0.1	0.18	0.24

a Are these readings precise? Explain your answer. ▶▶ **S50**

b State what these figures mean and explain what the river's grade should be. ▶▶ **WEA** **S14, S58, S59**

Long-Answer Grade Booster

★★★ clear explanation of percentiles

★★★ grade given using both oxygen and ammonia levels

★★★ grade given with reasons

A student was investigating the rate of reaction when calcium carbonate reacts with hydrochloric acid (to produce calcium chloride, water and carbon dioxide).

The student wanted to test the idea that the larger the surface area of reactants, the faster the reaction. She thought that if smaller pieces of calcium carbonate were used, then the reaction would go faster.

The student added hydrochloric acid to a conical flask and then added the calcium carbonate. The apparatus was on top of an electronic balance (which had a resolution of 1 decimal place). The mass was recorded every 10 seconds. The results are shown in the table.

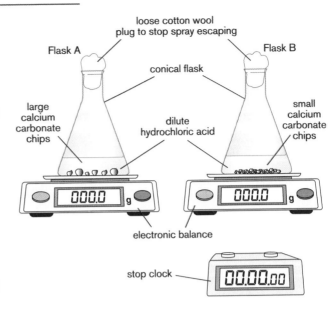

Figure A: A diagram of the apparatus used.

Time (s)	10	20	30	40	50	60	70	80	90	100	110	120
Mass of flask A (g)	52.7	52.0	51.5	51.1	50.8	50.5	50.3	50.1	50.0	49.9	49.9	49.9
Mass of flask B (g)	53.9	52.4	51.5	51.0	50.7	50.5	50.4	50.4	50.4	50.4	50.4	50.4

QUESTIONS

1 Write out a word equation for the reaction. ▶▶ **S31**

2 What would the student use to measure the volume of acid? ▶▶ **S44**

3 a What is the independent variable in this experiment?
   b What is the dependent variable?
   c State two control variables. ▶▶ **S43**

4 One chip in flask A had a volume of 0.52 cm³ and a surface area of 3.1 cm². What is its (surface area):(volume) ratio? ▶▶ **S32**

5 a Calculate the percentage loss in mass of the large chips. Give your answer to an appropriate number of significant figures.
   b Calculate the percentage loss in mass of the small chips. Give your answer to an appropriate number of significant figures. ▶▶ **S3, S5**

c Use your answers to parts a and b to explain how accurately you think that student performed this experiment. ▶▶ **S50, S58**

6 a What hypothesis was the student testing? ▶▶ **S40**
   b What prediction did the student make? ▶▶ **S40**
   c Explain whether the experiment supports the original hypothesis. ▶▶ **S34, S58**

7 Suggest two ways in which this experiment could be improved. ▶▶ **S43, S49, S58**

8 Calculate the rate of the reaction (in g of carbon dioxide per second) for each flask between 10 and 20 seconds. ▶▶ **S56**

9 a Draw an appropriate chart or graph to show both sets of results. ▶▶ **S23**
   b **HIGHER** Calculate the rate of the reaction in Flask B at 25 seconds. ▶▶ **S28**

The Thames Barrier shuts to stop the sea flooding London. Without it, London would be flooded once every 1000 years causing £80 billion of damage. The barrier halves this risk. However, global warming is raising sea levels. Since it became operational in 1982, the barrier shut four times in the 1980s, 35 times in the 1990s, 75 times between 2000 and 2009, and 61 times from 2010 to the end of 2015. By 2030, the barrier is not expected to reduce the flood risk below today's risk.

Increasing carbon dioxide (CO_2) levels in the atmosphere cause global warming. CO_2 absorbs heat and so stops the Earth losing so much heat into the atmosphere. Scientists collect data about CO_2 levels and global temperatures to create computer models that predict the effects of increasing CO_2 levels.

Figure A: The Thames Barrier cost £1.5 billion (at today's prices).

Figure B: CO_2 levels above a mountain in Hawaii.

To make predictions using the computer models, scientists need to forecast what will happen to the populations of different countries and how much energy those countries will need in the future. This is difficult to do and so they invent different 'scenarios' of what might happen in the future. Each scenario is then fed into the computer model. The chart on the left of Figure C shows the extra CO_2 that would be added to the atmosphere in different scenarios. The graph on the right then shows the predicted effects on the Earth's temperature for three of those scenarios (A2, A1B and B1).

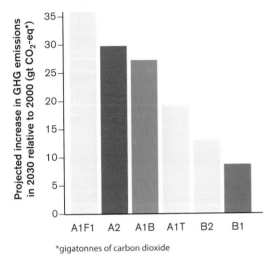

Extra CO_2 and other greenhouse gases in the atmosphere in 2030 compared with 2000.

*gigatonnes of carbon dioxide

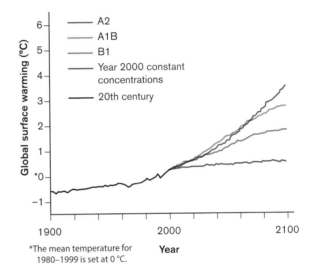

Yearly mean global surface increase in temperature for different scenarios (compared with the mean value for 1980–1999 (°C))

*The mean temperature for 1980–1999 is set at 0 °C.

Figure C: Computer model predictions of global temperature rises with different increases in CO_2.

QUESTIONS

1 Draw a graph or chart to display the times the Thames Barrier has shut. ▶▶ S23

2 Explain why two different types of chart/graph are used in Figure C. ▶▶ S9, S24, S58

3 Suggest what the term 'computer model' means. ▶▶ S31

4 a Scientists used Figure B to create a model for global warming. Suggest an advantage of using one set of data.
 b Suggest a disadvantage.
 c From where do you think CO_2 data should be collected for computer models? ▶▶ S12

5 Use Figure B to calculate the yearly rate of increase of CO_2 levels. ▶▶ S27

6 Calculate the percentage probability of London flooding:
 a today, without the Thames Barrier

 b today, with the Thames Barrier
 c in 2030, if there were no Thames Barrier ▶▶ S13

7 London's flood risk is quite low. Suggest why the Thames Barrier was built. ▶▶ S37, S58

8 Some people have suggested building a second Thames Barrier, at a cost of about £20 billion. Discuss this idea. ▶▶ WEA S38, S58, S59

Long-Answer Grade Booster

★★★ justifies views
★★★ describes how problems should be solved
★★★ identifies pros and cons

AIRBAGS

Airbags reduce the risk of injury in a car crash. One study of 2864 crashes discovered that 139 people suffered serious kidney injuries. Of these, 96 were in cars without airbags and 43 were in cars with airbags.

In a crash a sensor circuit produces a spark. In some airbag systems, the spark causes sodium azide (NaN_3) to decompose, releasing nitrogen gas. Sodium azide is very poisonous and can cause people to stop breathing. So, it is kept in a strong container.

Figure A: An airbag inflates in about 30 milliseconds.

$$2NaN_3 \rightarrow 2Na + 3N_2$$

An airbag needs a certain pressure of gas in it. It must not be too hard or it will hurt the driver but it won't work if it's too soft. Figure B shows the result of an investigation to find out what pressures were generated in different sizes of airbag.

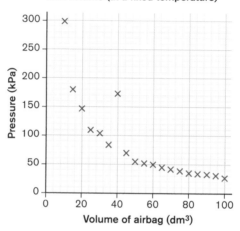

How the pressure of a gas-filled airbag varies with volume (at a fixed temperature)

Figure B: The pressure of gas produced by 130 g NaN_3 in different volumes of airbag.

QUESTIONS

1 How many seconds does it take for an airbag to inflate? ▶ **S55**

2 Write out the equation above in words, starting 'Two units of sodium azide...' ▶ **S18**

3 a A trial run was done for the pressure investigation. State three reasons why. ▶ **S45, S58**

   b State two control variables used. ▶ **S43**

4 a On the graph, which result would you check again and why? ▶ **S48**

   b What is the relationship between pressure and volume? ▶ **S25**

   c How would you use the data to draw a straight line graph? ▶ **S25**

5 130 g of NaN_3 is used for an airbag that needs 35 kPa of pressure. A cylinder-shaped bag is 26 cm tall and 58 cm in diameter. Is it the right size? Explain your answer. ▶ **S32**

6 Airbags are compulsory by law in some countries. Describe two factors that would be considered when deciding whether to make airbags compulsory in the UK. ▶ **S38, S58**

7 Discuss the fitting of airbags in cars. Think about the benefits, drawbacks and risks. ▶ **WEA S36, S47, S58, S59**

Long-Answer Grade Booster

★★★ risk reduction is evaluated

★★★ benefits, drawbacks and risk-reduction methods described

★★★ one benefit, drawback and risk-reduction method stated

C9 AVOGADRO'S BIG IDEA

Jöns Berzelius (1779–1848), an influential Swedish chemist, believed that particles combined using 'electrical force', so two oxygen atoms had the same charge and could not join in pairs.

In France, Joseph Gay-Lussac (1778–1850) showed that volumes of gases reacted in ratios of whole numbers. So, two volumes of hydrogen and one of oxygen reacted to form two volumes of water. This was difficult to understand (see Figure A).

A little known Italian, Amedeo Avogadro (1776–1856), then had some ideas:

* equal volumes of different gases contained the same number of 'particles'
* 'particles' could split (e.g. oxygen might have two joined particles).

Avogadro used his ideas to calculate the mass of oxygen particles compared to hydrogen particles. His ideas were ignored until Stanislao Cannizzaro (1826–1910) used them to accurately calculate the masses of atoms of different elements and did experiments to show that the calculations were correct. Avogadro's name is now used to describe 6.02×10^{23} particles as Avogadro's number or one 'mole'.

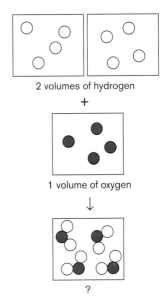

2 volumes of hydrogen
+
1 volume of oxygen
↓
?

Figure A: Chemists had trouble explaining Gay-Lussac's observations because if you joined up all the particles to make model water molecules you got only one volume of water, not two.

QUESTIONS

1 96 cm³ of hydrogen reacts with 32 cm³ nitrogen. What ratio is this? ▶▶ **S3**

2 Avogadro used 1.10359 for the density of oxygen and 0.07321 for hydrogen.
  a Give a unit used to measure density today. ▶▶ **S55, S56**
  b Write these figures to 3 significant figures. ▶▶ **S5**
  c From these figures Avogadro calculated how much heavier an oxygen particle was, compared to hydrogen (oxygen's 'relative atomic mass'). Calculate this figure. ▶▶ **S3, S58**
  d Use the information from the periodic table on page 126 to compare Avogadro's answer to today's figure. ▶▶ **S58**

3 Redraw Figure A to show what we now think happens in this reaction. ▶▶ **S31**

4 Draw a flow chart to show how Avogadro's ideas about how to calculate the masses of particles (compared with hydrogen particles) became a theory. ▶▶ **S33**

5 Suggest why Avogadro's ideas were ignored. ▶▶ **S39, S58**

6 (This type of question is only asked in single Physics GCSE.) At 'standard temperature and pressure' a mole of any gas always has the same volume. Use the information in the table to draw a chart/graph to calculate this volume of gas per mole. ▶▶ **S26, S27, S58**

Number of moles	2	4	6	8	10
Volume (dm³)	46	90	130	180	225

When athletes have sprains they often put a 'cold pack' on the affected area to reduce the swelling. Some cold packs use chemicals that cause energy to be taken out of the surroundings when they dissolve in water.

An experiment was done to test the idea that some chemicals are better than others for use in cold packs. 25 cm³ of water was added to a mixture of chemical powders and the drop in temperature was recorded. The balance could measure down to 0.1 g and the thermometer could measure down to 0.1 °C. The results are in the table. It was concluded that 5 g of sodium nitrate was best for making a cold pack.

Figure A: This cold pack contains ammonium chloride. The pack is twisted to release water into the chemical. As the ammonium chloride dissolves it becomes very cold.

Mass of ammonium chloride (g)	Mass of sodium nitrate (g)	Change in temperature (°C)		
		1st try	2nd try	3rd try
2.5	2	8.0	7.4	7.4
2.5	3	9.2	7.9	8.1
2.5	4	15.2	9.1	9.3
2.5	5	11.8	10.2	10.4
2.5	6	10.3	10.1	10.2

Figure B: Ammonium chloride is harmful if breathed in. Sodium nitrate is harmful if swallowed and irritates the skin.

QUESTIONS

1 What was the resolution of the thermometer?
▶▶ **S51**

2 Suggest two ways in which the risks of using these chemicals can be reduced.
▶▶ **S47, S58**

3 a Identify the most anomalous result.
▶▶ **S48, S58**

 b Calculate the mean temperature changes.
▶▶ **S6**

 c Draw an appropriate graph or chart to show the means. ▶▶ **S23**

 d What correlation does your graph/chart show? ▶▶ **S16**

4 Suggest two control variables for this investigation. ▶▶ **S43, S58**

5 Suggest what needs to be done to be more certain of the conclusion given. ▶▶ **S52, S58**

6 What hypothesis did this investigation want to test? ▶▶ **S40**

7 Evaluate this investigation.
▶▶ **WEA** **S54, S58, S59**

Long-Answer Grade Booster

★★★ justifies statements on the validity of the investigation

★★★ describes strengths and weaknesses of the investigation

★★★ comments on the quality of the data

Nuclear power uses heat produced by splitting atoms (nuclear fission) to generate electricity. It doesn't produce smoke or carbon dioxide but does produce radioactive waste (which can cause cancers).

Figure A: A nuclear explosion in Chernobyl in 1986 may have caused thousands of cancer cases.

Fusing atoms together produces heat but little radioactive waste. Fusion occurs in the Sun, at about 15 million °C. In 1989, scientists were trying to make fusion happen at room temperature. Stanley Pons and Martin Fleischmann were writing a paper about the experiment they had done. So was Steven E. Jones. They all agreed to send their two papers to *Nature* on the same day.

Pons and Fleischmann didn't wait for publication and gave a press conference, saying they could cause 'cold fusion' of deuterium atoms but didn't know how it worked. If deuterium atoms fuse, the accepted theory says that heat and helium gas are produced, along with neutrons, protons or gamma rays. Pons and Fleischmann only measured heat increase and there was no control experiment. Other scientists could not replicate their work nor detect helium, protons, neutrons or gamma rays.

Palladium is a metal. They used different volumes of palladium in their experiments.

platinum   palladium

D_2O

dc power supply

If the deuterium atoms fuse in the palladium, heat will be released and the temperature will rise. The heat energy will be more than the electrical energy put into the device.

OD^-

Deuterium is a type of 'heavy hydrogen', which can be used to make 'heavy water' (D_2O). They added different amounts of this to the apparatus during their experiments.

When electricity is passed through the solution, deuterium is released from the D_2O and is absorbed by the palladium.

Figure B: Pons and Fleischmann's experiment.

QUESTIONS

1 Suggest a benefit and a drawback of nuclear power based on fission. ▶▶ **S36, S58**

2 a What hypothesis did Pons and Fleischmann test? ▶▶ **S40**

 b Suggest two predictions made by this hypothesis. ▶▶ **S40, S58**

3 Write out 15 million °C in standard form. ▶▶ **S2**

4 Explain why a paper should be published before press conferences are given. ▶▶ **S39, S58**

5 a Suggest how Pons and Fleischmann acted unethically. ▶▶ **S38, S58**

 b Give two reasons why most scientists don't believe their results. ▶▶ **S33, S40, S46, S49**

6 Pons and Fleischmann may have over-estimated the amount of heat energy they produced because they didn't stir the liquid. Suggest how. ▶▶ **S52, S58**

7 Explain what a control is and suggest some controls for Pons and Fleischmann's experiment. ▶▶ **WEA S46, S58, S59**

Long-Answer Grade Booster

★★★ explains why a control is needed, with correct suggestions

★★★ describes what a control is, with a correct suggestion

★★★ states the need for a control, with a suggestion

According to legend, Galileo Galilei (1564–1642) dropped two masses from the top of the leaning tower of Pisa. One mass was ten times heavier than the other but they hit the ground (56 m below) at the same time.

The legendary experiment tested Galileo's idea that gravity acts on all objects equally. This means that all dropped objects accelerate towards the Earth at the same rate. Galileo's results showed that Aristotle (384 BCE–322 BCE) was wrong when he claimed that a mass that was ten times heavier would fall ten times faster.

Figure A: Historians doubt whether Galileo himself dropped masses from the tower.

However, some objects are slowed by air. There is no air on the Moon and so in 1971 an astronaut dropped a hammer and a feather at the same time. Both objects hit the ground together.

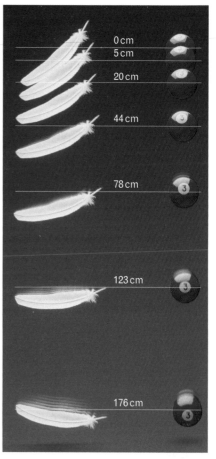

0 cm
5 cm
20 cm
44 cm
78 cm
123 cm
176 cm

Figure B: A feather and a ball dropped at the same time in a vacuum (no air at all), with an image taken every 0.1 seconds.

QUESTIONS

1 Is the data from Galileo's experiment quantitative or qualitative? ▶▶ **S41**

2 Convert 0.1 seconds into milliseconds. ▶▶ **S21, S55**

3 Use the word 'reproducible' to explain why we think Galileo's idea is correct. ▶▶ **S49, S58**

4 a Calculate the mean speed of the ball in Figure B. ▶▶ **S56, S58**

  b Suggest why Galileo was not able to calculate the objects' speeds. ▶▶ **S51, S58**

5 a Calculate time squared (t^2) for each position of the ball in Figure B. Present your answer as a table. ▶▶ **S7, S57**

  b Draw a scatter graph with a line of best fit for distance against t^2. ▶▶ **S15**

c Describe the relationship shown on your graph. ▶▶ **S15, S16, S58**

d Use the information in your graph to calculate the constant of proportionality. ▶▶ **S27**

6 Calculate the maximum speed of an object dropped from the top of the tower. Use:

$$v^2 - u^2 = 2 \times a \times x$$

(a = the acceleration due to gravity, which is 9.8 m/s²) ▶▶ **S19, S20, S22**

7 Draw a flow chart to show how Galileo's idea has become a theory. ▶▶ **S33**

P3 PET SCANS

In the 1920s, Paul Dirac (1902–1984) developed a mathematical equation to model what happens to negatively charged electrons in different situations. His equation only worked if electrons with positive charges also existed. So, he predicted 'positrons', which were found in 1932 and are used in PET scanners.

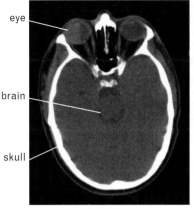

A CT scan

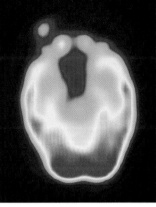

A PET scan

Figure A: Different scans of a person's head. The scans are from a study to work out how good PET scanners are at detecting eye cancers.

In a PET scan, a radioactive version of a compound that your body uses (such as glucose) is injected. The compound, called a tracer, produces positrons. The scanner detects where the positrons are produced and displays the results on a screen.

Tracers are very expensive and need to be made close-by, so only some hospitals have PET scanners. A scan increases the risk of cancer by 1 in 10 000. The risk is greater if you are still growing, so children and pregnant woman don't usually have PET scans.

QUESTIONS

1 a What scientific question did Paul Dirac ask?
  ▶▶ **S35**

 b What prediction did his model make?
  ▶▶ **S40**

2 The table shows more results from the study into PET scans and eye cancers.

Patient number	Average radius of cancer (mm)	PET scan detected the cancer?
1	4.63	No
2	5.25	No
3	8.00	Yes
4	5.50	No
5	9.25	Yes
6	9.00	Yes
7	7.75	Yes
8	8.00	Yes
9	6.25	Yes
10	3.25	No

 a Assume that a cancer is a sphere. Calculate the volume of the cancer in each patient. Give your answers to 3 significant figures.
  ▶▶ **S5, S32, S57**

 b Why are the volumes you calculated only estimates? ▶▶ **S4, S31**

 c This was a small study, like a trial run. Explain one reason why scientists use trial runs. ▶▶ **S45, S58**

 d What range of cancer radii would you choose to test to work out a better value for the resolution of a PET scanner in detecting eye cancers? ▶▶ **S6, S51**

3 Discuss the use of PET scans. Consider the benefits, drawbacks and risks.
  ▶▶ **WEA** **S3, S36, S47, S58, S59, S60**

Long-Answer Grade Booster

★★★ benefits, drawbacks and risks explained using examples

★★★ benefits, drawbacks and risks described

★★★ one benefit, drawback and risk stated

Ig® Nobel Prizes are given to scientists whose research makes people laugh and then makes them think. Dr Lianne Parkin won a prize in 2010 for investigating the effect of wearing socks outside shoes in icy conditions.

Figure A: Dr Parkin receiving her award from Prof. Sheldon Glashow.

Dr Parkin and her team found 30 volunteers on an icy footpath, and gave each an envelope at random. Half the envelopes contained the word 'socks' and the other half contained the phrase 'no socks'. Those in the 'socks' group were given 'acrylic-blend work socks' to put on, *over* their footwear. The people were timed as they walked down the path and asked how slippery they thought the path was on a scale of 1–5.

One person didn't complete the task. One person had a sock flapping off her shoe, creating a hazard, but her results were included. During planning, Dr Parkin thought about using the world's steepest road but decided this would be 'ethically … unwise'.

Group	Number of ...		Age range	Slipperiness ratings (5 is most slippery)	Times taken to descend the slope (s)
	men	women			
Socks	7	7	19–58	1, 1, 1, 1, 1.5, 2, 1.5, 1, 1, 1, 1, 3, 1, 5	23.9, 27.2, 32.0, 42.4, 38.9, 29.2, 45.0, 26.1, 40.9, 43.0, 48.3, 31.9, 55.0, 44.0
No socks	10	5	18–70	1.5, 2, 2, 1, 1.5, 2, 2.5, 4, 4, 3, 4, 4, 5, 5, 2	29.2, 32.0, 35.4, 48.0, 29.0, 27.8, 30.8, 44.4, 48.4, 48.4, 29.9, 69.4, 39.7, 49.0, 33.1

QUESTIONS

1 Give one variable that was controlled and one that was not. ▶▶ **S43**

2 a Which group was the control? Explain your reasoning. ▶▶ **S46, S58**

b Why was a control group used? ▶▶ **S46**

3 What is the most anomalous reading in the times taken to descend? ▶▶ **S48**

4 a Calculate means for the slipperiness ratings and times for both groups. Use all the results. ▶▶ **S6, S58**

b Why are means calculated? ▶▶ **S6**

c Show the means on two appropriate graphs or charts. ▶▶ **S23**

5 Explain why using the steep street would have been 'ethically unwise'. ▶▶ **S38, S58**

6 Draw a tally chart to show the slipperiness ratings given by the people who wore the socks. ▶▶ **S8**

7 Why is it important that the groups were chosen randomly? ▶▶ **S12**

8 Suggest another question that could be answered by a similar investigation. ▶▶ **S35, S58**

9 Identify the hazards mentioned on this page and explain how the risks should be reduced, producing an argument in support of your ideas. ▶▶ **WEA** **S47, S58, S59, S60**

Long-Answer Grade Booster

★★★ explains evidence for risk reduction as part of an argument

★★★ correctly refers to hazards, risks and risk reduction

★★★ identifies a danger

High-pitch deterrent devices (HPDDs) emit high-pitched sounds that only younger people can hear. They can stop groups of young people loitering.

You hear high-frequency sound waves as high-pitched sounds. The frequency of a wave is measured in hertz (Hz). Loudness can be measured in decibels (dB).

A shopkeeper wanted to find the best pitch and loudness settings for an HPDD. Lucy (aged 14) and her dad Brian (aged 47) agreed to help. They listened to different frequencies. The loudness of each frequency was increased until they heard the sounds.

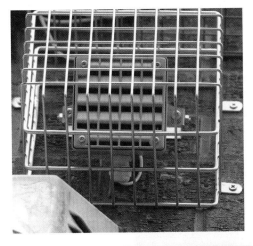

Figure A: An HPDD.

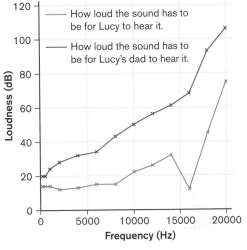

Figure B: The shopkeeper's results.

QUESTIONS

1 What is 12 000 Hz in kilohertz? ▶ **S55**

2 Which measurement needs checking? Explain your choice. ▶ **S48**

3 What correlation does the graph show? ▶ **S16**

4 The loudest sound wave Brian heard had a pressure of 2 000 000 μPa.
  a What does the μPa mean? ▶ **S55, S56**
  b Why do scientists use symbols? ▶ **S18**
  c Convert this figure into standard form. ▶ **S2**
  d The loudest sound wave Lucy heard had a pressure of 5×10^3 μPa. In μPa, how much less was this pressure compared to the one above? Show your working in standard form. ▶ **S2**

5 a An HPDD can be set to 8 or 17.4 kHz, and 40, 60 or 100 dB. Explain which settings you would recommend. ▶ **S55, S58**
  b Should the shopkeeper have used secondary data? Explain your reasoning. ▶ **S42, S58**

6 Brian said "The experiment shows that adults can't hear sound waves over 25 000 Hz." Is this valid? Explain your reasoning. ▶ **S52, S53, S58**

7 A shop is thinking about installing an HPDD to stop young people hanging around outside it. However, some people have said it is wrong to use a device that affects all young people. Explain how you would go about deciding for or against installing an HPDD. ▶ **WEA** **S38, S58, S59**

Long-Answer Grade Booster

★★★ explains the full range of evidence needed

★★★ describes some other evidence needed

★★★ states some simple factors to take into account

Distances in space can be measured in degrees (like angles). Such a measurement is the number of degrees you turn when moving from looking at one point to another. 1/60th of a degree is called an arcminute and an arcsecond is 1/60th of an arcminute. Modern telescopes can measure fractions of arcseconds.

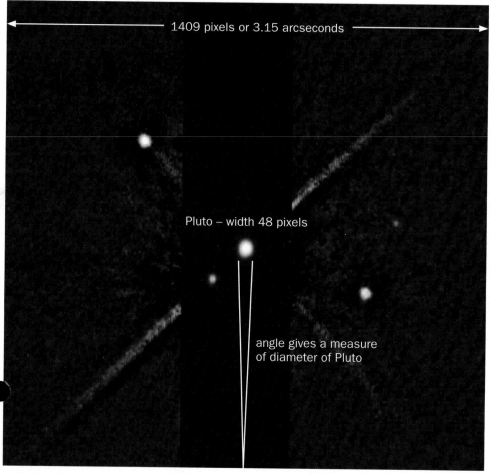

Figure A: Pluto as seen using a telescope in America. Its diameter can be expressed in arcseconds.

The smallest measureable distance in the 16th century was about half an arcminute. At this time Nicolaus Copernicus (1473–1543) proposed that the Earth went around the Sun, not the other way round. This explained why, during a year, planets got brighter and dimmer and why some planets appeared to travel backwards.

Copernicus' idea predicted parallax for some stars (see Figure B). No one could observe this and so many thought that Copernicus was wrong. A star's parallax was first observed in 1838 (as an angle of 0.314 arcseconds).

Viewpoint A. Looking at the star it appears in a certain position against the background of stars that are even further away.

Where the star appears from viewpoint B.

This angle is called the parallax angle.

Where the star appears from viewpoint A.

Viewpoint B. The Earth has now moved on its orbit around the Sun and so the star appears in different place against the background of other stars.

Not to scale

Figure B: Parallax would not happen if the Earth didn't move position.

QUESTIONS

1 a What 16th century evidence supported Copernicus' hypothesis? ▶▶ **S40**
  b Explain how Copernicus' hypothesis became a theory. ▶▶ **S34, S58**

2 Why wasn't parallax of a star observed by Copernicus? ▶▶ **S51**

3 How much of a degree is one arcsecond? Show your working. ▶▶ **S3**

4 a Use the information in Figure A to calculate the width of Pluto in arcseconds.
  b Show your answer as a fraction.
  c Give your answer to part a in arcminutes. ▶▶ **S3, S58**

5 a The table shows parallax angles of various stars and their distances from the Sun. Plot the data on an appropriate chart or graph. ▶▶ **S26**
  b **HIGHER** How are the two variables related? ▶▶ **S25**

6 a **HIGHER** Parallaxes are now measured by satellites. Before this, all parallaxes had been overestimated. What effect would this have on the distances? ▶▶ **S25, S50**
  b What sort of error was this? ▶▶ **S51**

7 Measure the parallax angle in Figure B. ▶▶ **S30**

8 Explain how the observations on this page confirmed Copernicus' theory. ▶▶ **WEA** **S39, S58, S59**

Long-Answer Grade Booster

★★★ explains expected observations if Copernicus was wrong

★★★ describes observations using Copernicus' theory

★★★ identifies some observations

Parallax (arcseconds)	0.77	0.55	0.37	0.27	0.23	0.22	0.21	0.20
Distance (light-years)	4.2	6.0	8.7	12.0	14.3	15.1	15.8	15.9

Sunshine is good for you because its ultraviolet (UV) rays help your skin make vitamin D (also found in fish and eggs). However, you should use a high SPF sunscreen if in the sun for some time. The SPF is the ratio of how long you can be in the sun wearing sunscreen before burning, compared to without sunscreen.

Figure A: Sunscreens display SPF (sun protection factor) numbers.

Sunscreens protect you against ultraviolet B rays, which cause sunburn. UVA rays are not stopped by all sunscreens but can damage cells, even though they don't cause sunburn.

An investigation was done to see if high SPF sunscreens protect skin better than low ones. A type of white bead goes purple when exposed to UVB rays. 21 beads were placed on a card and covered with colourless plastic film. Different sunscreens were smeared on the film over each bead. The card was then placed under a UVB lamp. The results are shown in the table.

SPF	Time taken to go purple (s)		
	Bead 1	Bead 2	Bead 3
4	70	75	74
8	131	134	134
15	276	278	274
30	688	542	798
50	1949	1900	1896

QUESTIONS

1 a How can you reduce the risk of sunburn? ▶▶ **S47**

b People are more likely to use sunscreens if warned about cancer than if warned about sunburn. Suggest why. ▶▶ **S37, S58**

2 a What is the independent (input) variable in the investigation?

b State two variables that should have been controlled. ▶▶ **S43**

3 Jack can have 10 minutes in the sun before he starts to burn. How long could he have when using an SPF 15 sunscreen? ▶▶ **S3**

4 Which set of readings in the table is the least precise? ▶▶ **S7, S50**

5 a Calculate the means for each set of readings in the table. ▶▶ **S6, S58**

b Plot your means on an appropriate chart or graph. ▶▶ **S23**

c What correlation does your graph show? ▶▶ **S16**

6 List points that could be made as part of an evaluation of the investigation. ▶▶ **S54, S58**

7 Using this page and Activity B9 (page 62), explain whether or not you think sunbathing is a good idea. ▶▶ **WEA** **S58, S59, S60**

Long-Answer Grade Booster

★★★ synthesises information

★★★ uses a lot of information from both pages

★★★ states a reason for and against sunbathing

'Average speed cameras' enforce a speed limit of 50 mph (80 km/h) at motorway roadworks. Two cameras, a certain distance apart, take photos marked with times of every car. A computer calculates the mean speed for each car using speed = distance ÷ time.

The speed limit on most motorways is normally 70 mph (113 km/h) but some people want this to be higher.

HIGHER They claim that the limit should be set using the speed that the 85th percentile of drivers would drive at on a motorway (about 80 mph).

Others disagree and say that raising the limit would encourage people to drive faster, increasing the risks of serious injury (see the table).

Figure A: Average speed cameras.

Speed change after impact (mph)	1–10	11–20	21–30	31–40	41–50	50+
Probability of serious injury	0.01	0.026	0.111	0.279	0.406	0.543

Data from a 1994 study

Figure B: Graph used by supporters of raising the speed limit based on studies carried out in the 1950s and 60s.

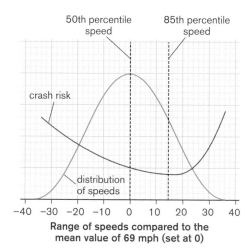

Range of speeds compared to the mean value of 69 mph (set at 0)

QUESTIONS

1 a Which units for speed on this page are SI units? ▶▶ **S55**

b The units for speed are compound units. What does this mean? ▶▶ **S56**

2 a The times on two photos from speed cameras seven miles apart for one car are 14:03:02, 14:10:4 (hours:minutes:seconds). How long in seconds did the car take? ▶▶ **S21, S55**

b What was the car's speed in mph? ▶▶ **S21**

c Cars going over a certain mean speed are sent a fine. This speed is 10% over the speed limit + 2 mph. Will the car driver be fined? ▶▶ **S3**

3 A car travels at 40 km/h for 15 minutes. How far will it go? ▶▶ **S19**

4 What is the percentage risk of serious injury if the change in speed after impact is 31–40 mph? ▶▶ **S13, S37**

5 a **HIGHER** What is meant by the 85th percentile?

b Where on the graph would you find the median value? ▶▶ **S14**

6 Display the data in the table on a suitable graph or chart. ▶▶ **S23**

7 Discuss the value of speed limits on motorways in the UK. ▶▶ **WEA** **S38, S58, S59**

Long-Answer Grade Booster

★★★ indication of what additional evidence there should be

★★★ point of view backed up with evidence

★★★ identification of the pros and cons of speed limits from the information given

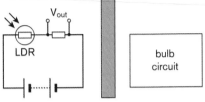

Dan wanted to see if the amount of light reflected from a board could be used to measure its distance accurately. To detect the light, he had to decide between using a light dependent resistor (LDR) or a photodiode. During pre-tests he found that the LDR was more sensitive. He set up two circuits; one containing an LDR and one with a switch and a bulb. The circuits were separated by a piece of wood.

When the white board was 1 m away from the bulb, he measured the voltage across the resistor in the LDR circuit. He measured the voltage at other distances and plotted Figure B.

He then placed the white board at other distances. He measured the voltage and used this to predict the distance of the board, before using a ruler to measure the actual distance. His results are shown in the table. Dan concluded that measuring distances like this was possible but not very accurate.

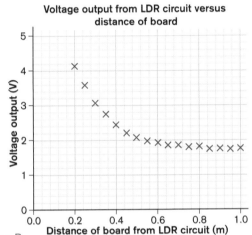

Figure A: Dan's set up.

Figure B

Position	Distance predicted by Figure B (m)	Actual distance (m)
1	0.22	0.21
2	0.46	0.43
3	0.75	0.79

QUESTIONS

1 a What trial run did Dan do? ▶▶ **S45**
  b State two reasons for doing a trial run.
    ▶▶ **S45, S58**

2 What range did Dan choose for the independent variable? ▶▶ **S6, S43**

3 The 'bulb circuit' had a battery, bulb and switch. Draw a circuit diagram. ▶▶ **S18**

4 a Use the graph in Figure B to interpolate the voltage when the board was at a distance of 0.26 m. ▶▶ **S15**
  b At this distance, the current in the circuit was 0.1 A. Calculate the resistance of the resistor. ▶▶ **S20**
  c At this distance, calculate the power output of the resistor. ▶▶ **S20**

5 What is meant by 'the LDR was more sensitive' than the photodiode? ▶▶ **S51**

6 When the board was in 'Position 2', what would the voltage have been? ▶▶ **S15**

7 a Calculate the percentage error for each predicted distance. This is the difference between the predicted and actual values compared with the actual value.
    ▶▶ **S3, S58**
  b What sort of error is this? ▶▶ **S51**
  c Suggest one way in which this error might have been caused. ▶▶ **S51, S58**

8 Evaluate this investigation.
    ▶▶ **WEA** **S54, S58, S59**

Long-Answer Grade Booster

★★★ justifies statements on the validity of the investigation

★★★ describes strengths and weaknesses of the investigation

★★★ comments on the quality of the data

A student was investigating the length of a spring as different masses were added to it. The results are shown in the table.

Mass added (g)	Length (cm) 1st try	Length (cm) 2nd try	Length (cm) 3rd try
0	9.8	9.8	9.8
50	19.3	19.3	19.3
100	29.2	29.3	29.3
150	39.2	39.1	39.1
200	49.1	49.0	49.1
250	58.7	58.7	58.8
300	68.5	68.4	68.3
350	78.2	78.2	78.5
400	87.6	87.6	87.9
450	97.2	97.1	97.3

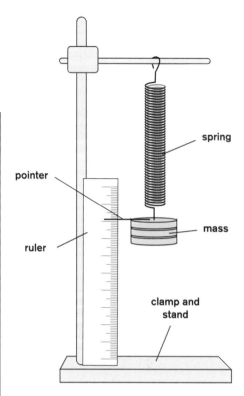

Figure A: A diagram of the apparatus used.

QUESTIONS

1 State the extension of the spring when 50 g was added, giving your answer in:
  a metres
  b mm. ▶ **S21**

2 State and explain which set of readings is the most precise. ▶ **S50, S58**

3 a Suggest a reason for the small amounts of random error shown in the results.
  b Explain how a systematic error could be caused in this experiment. ▶ **S51, S58**

4 a Calculate the mean lengths for each mass, showing your answers in a table. ▶ **S5, S6, S7**
  b State the mean of the readings when 450 g was added, showing the degree of uncertainty. ▶ **S5, S51**
  c Plot your mean values on an appropriate graph. ▶ **S23, S24, S26**
  d State the intercept of your line. ▶ **S27**
  e Calculate the gradient of the line. Show your working. ▶ **S27**
  f State the equation for your line. ▶ **S27**

5 a The length of the spring plotted against the mass shows a linear relationship. By plotting the extension of the spring against mass, it can be shown that the extension is directly proportional to the mass applied. What does this mean? ▶ **S25, S27**
  b What symbol is used to show direct proportion between two variables? ▶ **S18, S27**
  c Extension can also be plotted against force (in newtons). Calculate the force due to gravity on a 50 g mass. (g = 9.8 N/kg) ▶ **S20**

6 **HIGHER** The area under the line on a graph of extension plotted against force is the work done in changing the extension of the spring, in newton metres (N m).
  a Explain how you would calculate the area under the line. ▶ **S29, S58**
  b Use the results in the table to draw a graph to calculate the work done in changing the extension of this spring. Show all your working. ▶ **S20, S21, S24, S29**

Answers

There are answers to all the questions in this section. There are many different ways in which questions are asked, so have a look at S58 if some of the command words (the words at the start of a question telling you what to do) are a bit confusing.

Some of the answers have additional notes in *italics* next to them. These notes make general comments about a question and describe some common errors made in answering questions of a particular type.

🏠 Some answers have this icon. This means that there is a better way of answering the question, which you might have missed. In an exam, questions are given a number of marks. Sometimes a question might just have 1 mark and the basic answer will get you that mark. However, that same question may be worth 2 marks, in which case you'd need the 'better answer'. It's always good practice to write the best answers that you can!

WEA In exams, some questions require a long written answer and may be worth up to 6 marks. ('WEA' stands for Writing Extended Answers.) Your answer should be well structured, and you should also try to use good spelling, punctuation and grammar. Sections S58, S59 and S60 contain some information about how to answer long-answer questions.

WEA In a 6-mark, long-answer question you'll get **1–2 marks** if you:
- ✖ write down basic scientific information
- ✖ show a simple understanding of the science
- ✖ don't use many scientific words (or show a lack of understanding of their meaning)
- ✖ don't organise your answer
- ✖ don't include much detail.

You'll get **3–4 marks** if you:
- ✖ write down accurate scientific information
- ✖ show a clear understanding of the science
- ✖ use some scientific words, some of which are not used properly
- ✖ try to organise your answer
- ✖ include some detail.

You'll get **5–6 marks** if you:
- ✖ write down scientific information that is always accurate and relevant
- ✖ show a detailed understanding of the science, which you demonstrate by giving evidence and examples
- ✖ use a wide range of relevant scientific words, which are all used properly
- ✖ organise your answer so that it forms a logical series of points
- ✖ include detail.

In the answers to these questions, there is a list of points that you could have included in your answer. The points marked ★ are those that would be found in a 1–2 mark answer. The points marked ★★ are those that would be found in a 3–4 mark answer. The points marked ★★★ are those that would be found in a 5–6 mark answer.

On the following pages, two of the long-answer questions have been answered and comments have been made on the different answers. Have a look at these to help you write better longer answers.

6 Some blood cells are put in pure water. Explain what will happen.
▶▶ **WEA** S40, S58, S59

★ **Low level answer**

There are lots of misspellings in this answer. Always try to use the correct spellings of scientific words.

Things can move into the sell and make changed the shape this is because water molickules can move. The sells get all big and fat.

This is poor grammar. It should read 'and change the shape'. Try to use good grammar to ensure your meaning is understood.

This is poor punctuation. Sentences should end with a full stop and start with a capital letter, otherwise it is very difficult to read. This should read '... the shape. This is because ...' .

The way in which this answer is written is not very logical. Answers need to be written in a way that lets the reader follow a process in logical steps. It would make much more sense if it were written like this:

Things can move into cells and change their shape, making them big and fat. This is due to the movement of water molecules.

Two scientific points are made, however, and these would gain a mark or two.

★★ **Medium level answer**

This is a common spelling mistake. Always try to use the correct spellings of scientific words.

This is not correct. The moving of substances into and out of a cell is not osmosis. This term is only used to describe water (or other solvent) moving through a selectively permeable membrane. Also, the position of that sentence is not great because it's not clear whether the student means that the process of substances moving is called osmosis or whether when a cell changes its shape it's called osmosis.

A sell can change it's shape when substances move into or out of it. This is called osmosis. If a blood sell is put in pure water, water molecule's move into it making it swell up.

This is poor grammar. The first line should read 'its shape'. When you are saying 'it is' then you can turn that into 'it's'. However, if you are saying that something belongs to 'it' then there is no apostrophe. This is a very common grammatical error. In the last sentence there is an apostrophe in molecules. This is also incorrect but is quite a common error. Plural nouns do not have apostrophes.

The punctuation in this answer is good.

There are a few scientific points that are made and presented in a fairly logical order and marks will be awarded for this. However, the question asks you to 'explain' how the process happens and this has not been done.

★★★ High level answer

This is excellent and makes good use of scientific words like 'membrane' and 'selectively permeable'.

There is a very clear explanation of osmosis, which is what the question asked for.

The membrane of the blood cell is selectively permeable, which means that only water molecules can pass through it. If there are more water molecules on one side of the membrane than the other, there will be more water molecules that move through the membrane to the side where there are less of them (a process called osmosis). Fewer water molecules will move in the opposite direction. When a blood cell is put in pure water, there are more water molecules outside the cell and so there will be an overall movement of water molecules into the cell by osmosis. This will make the cell swell up.

This answer has a very good structure, giving a clear explanation of the process of osmosis before going back to the example in the question to show how the process applies to that example. This answer would get full marks.

P3 PET SCANS – worked example

3 Discuss the use of PET scans. Consider the benefits, drawbacks and risks.
▶▶ WEA S3, S36, S47, S58, S59, S60

★ Low level answer

This not spelt correctly. PET is an abbreviation (for Positron Emission Tomography but you don't need to know that) and so is spelt with capital letters. It doesn't look good if you misspell words that are in the question.

Pet scans are good because they find cancer, which can kill people. They save a lot of lives because they finding lots of different sorts of cancer like stomack cancer lung cancer kidney cancer and brain cancer. This means that people with cancer can get treated because they will know that they have cancer. This save there lives. All hospitals should have a pet scanner.

This poor punctuation makes the answer difficult to read. Lists should have commas to separate out the different items. Also this sentence is just a space filler. Putting in a list of different cancers does not answer the question.

This answer is not good. It does not answer the question. It may score 1 mark (for stating that the student is in favour of something and giving a reason why). It also contains many spelling, punctuation and grammatical mistakes, which makes it difficult to read.

When you are asked to discuss something, you need to build up an argument (as shown in S60). There is no argument built here.

★★ Medium level answer

These are good reasons against using PET scanners.

PET scans can detect some cancers that other methods cannot detect. They are also quick and so paysents can get their results quickly and start their treatment quickly.

However, PET scanners are very expensive and radioactive supstances are used, which can be dangerous.

So overall I think that PET scanners are good things to have in hospitals.

Make sure you check your spelling, punctuation and grammar in your answers to longer questions, to help the examiner understand the points you are making.

This answer communicates what the student thinks. However, an argument has not been developed, the student has simply stated some reasons why PET scanners are a good thing and some reasons why they are not. An argument needs to have some reasons behind these statements.

★★★ High level answer

This is a clearly stated point of view. Remember that such an opinion is not necessarily right or wrong.

This is a good structure, with three clear paragraphs setting out and explaining the reasons why the student holds that opinion.

This is a clear counterargument, followed by a response saying why the student doesn't agree with the counterargument.

This is a very clear argument, with good spelling, grammar and punctuation that make it easy to follow.

I do not think that PET scans should be used in the UK.

PET scanners are very expensive, which uses up a great deal of money that would be better spent on treating more people.

Not every hospital can have a PET scanner because they need to be near a place where the radioactive tracers are made. This means that people may have to travel a long way to get a PET scan. It would be better to have cheaper alternatives in more hospitals.

Dangerous radioactivity is used, which can itself cause cancer. Although the risk of getting cancer from a PET scan is small it's large enough to mean that children and pregnant women aren't given PET scans. So why should other people be subjected to these scans?

Some people will say that PET scans allow smaller cancers to be found. However, there are plenty of other ways of detecting cancer and we should invest the money saved from PET scans into making these better.

The dangers and expense involved with PET scans means that we should stop using them.

1 An estimate is a rough calculation. *This is a straightforward question.*

2 **a** $5 \times 4 = 20\,km^2$. *Don't forget the units. If there were marks for this question, it would be out of 2 – one for the correct calculation and one for the units.*

   b 140/20 = 7. The whole area is 7 times bigger than the sample area. So, there are about 7 × 51 white rhinos in the whole area = 353. 🔺 A better answer would be to round this down to 2 significant figures, which is 350. The answer is an estimate and the two area measurements are given to 2 significant figures. *As a rule, give your answers to the same number of significant figures as most of the numbers in the question are given.*

3 14 500:100 = 145:1 *You need to cancel the ratio down, as you would a fraction, in order to get it into its simplest form.*

4 5/20 = ¼ or 0.25 or 25%. *You have three choices of how you express your answer!*

5 **Proportions of the different species of rhinoceros**

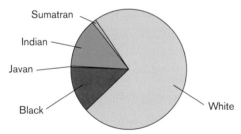

You've been asked to draw a pie chart for this data because you are showing the different proportions that contribute to a whole.

The angles you need are calculated as in the table below.

Make sure you have:
- *the correct angles*
- *a title*
- *all the categories neatly labelled*
- *used a ruler to draw the lines*
- *used a compass or similar to draw the circle.*

6 The equipment needs to be checked to make sure it works. Specifically, the automatic cameras need to be tried with real animals (or maybe Sumatran rhinos in captivity) to make sure that they do take photos when an animal walks past.

7 **WEA** *Check your spelling, grammar and punctuation and try to include scientific words, such as 'sample' and 'estimate'.*

You should try to include some of the points in the list below. You don't need to make all the points but you should aim to make points with the more stars. Also look at the grade booster box on page 53 and the general points about writing extended answers on page 86.

★ you would need to count the numbers of white rhinos in different years

★ if the numbers are increasing or staying about the same, conservation efforts are working

★ you count the number of rhinos in a sample area of known size

★ this allows you to calculate an estimate of the number of rhinos

★ you multiply the number of rhinos counted by how many times bigger the whole area is compared with the sample area

★★ you don't need an exact number to know if generally the number of rhinos is increasing or decreasing

★★ there are too many white rhinos to count each one

★★★ counting rhinos individually would take too long/be too expensive

★★★ it would be very hard to get an exact number because rhinos will be being born and dying whilst the count is going on

To get the most marks for a question like this, you need to justify your answer. This means explaining why the way in which something is done is the best way of doing it.

Rhino species	Sumatran	White	Black	Javan	Indian	Total
Number	100	14 500	2500	60	2650	19 810
Angle calculation 360/19 810 = 0.0182	100 × 0.0182	14 500 × 0.0182	2500 × 0.0182	60 × 0.0182	2650 × 0.0182	
Angle	2°	264°	45°	1°	48°	360°

B2 SALTY SNACKS

1 $1/100\,000 \times 100 = 0.001\%$

2 Qualitative – there are no numbers.

3 A membrane that allows only some substances (e.g. water) to pass through it. *This question requires you to have understood how this sort of membrane works (from the text) and to have examined Figure C in enough detail to find out what a membrane like this is called.*

4 To help them to understand how something works. (They can also be used to make predictions but you won't have got that from this page.) *This is a straightforward question to make sure you understand what a model is.*

5 a There are two points here. If this were a 1 mark question you would only need one point, if it were a 2 mark question you would need both points:
- because using a single cell may give an anomalous result/outlier
- the more measurements you make, the more sure you can be of what you are observing.

b Using pure water. *A control is usually a repeat of the experiment in which the independent (input) variable is not applied.*

c Two from: the type of salt; the type of cell; the temperature; the apparatus used; the volume of salt solution; the number of cells.

6 **WEA** *There are some examples of answers to this question, and comments on those answers, on pages 87–88.*

If a question asks you to explain something then take great care that you <u>explain how</u> something is happening, and you don't just <u>describe what</u> happens. You could answer this question more clearly by using a model. You could draw a version of Figure C as part of your answer.

Check your spelling, grammar and punctuation and try to include scientific words, such as 'dilute', 'concentrated', 'osmosis', 'membrane' and 'molecule'.

You should try to include some of the points in the following list. You don't need to make all the points but you should aim to make points with the more stars. Also look at the grade booster box on page 54 and the general points about writing extended answers on page 86.

★ the cells will change shape
★ substances can move into and out of the cell
★★ osmosis occurs
★★ water moves into the cells
★★ the cells swell up/become rounder
★★★ water molecules move from a place where there are more of them (more dilute solution) to a place where there are less of them (more concentrated solution)
★★★ other molecules of dissolved substances cannot go through the membrane, so only the water can move
★★★ the membrane is selectively permeable
★★★ more water molecules move into the cell than out of the cell (another way of saying this is that there is 'a net movement' of water molecules into the cells)

B3 DAR-WINS!

1 A theory is a scientific idea that is supported by a lot of evidence. A hypothesis is a scientific idea that has little or no evidence to support it (but no evidence against it).

2 Reading Malthus' essay. *Remember to read the question carefully. This question is about how evolution works and not about what evolution is (gradual changes in organisms).*

3 *Hyracotherium* is one order of magnitude older than *Equus*.

4 a secondary

b Two of: you can't control the quality of the data; the scientists who find different fossils may interpret them in different ways; the scientists who find different fossils may dig them up using different techniques; the drawings/photographs of the fossils may not be very good; the drawings/photographs of the fossils may not show the features you are interested in.

5 35 million years ago is $35\,000\,000 = 3.5 \times 10^7$ years ago.

6 It may cause bias. The fossil may be the only one that shows a certain feature whereas all the other members of this species did not have that feature at the time.

7 A line graph, to show changes with time.

8 [WEA] *You need to look at all the information that you've been given on the page. When writing a long answer like this, you need to give relevant examples as part of your explanation.*

Check your spelling, grammar and punctuation and try to include scientific words, such as 'evolution', 'theory', 'evidence', and 'species'.

You should try to include some of the points in the list below. You don't need to make all the points but you should aim to make points with the more stars. Also look at the grade booster box on page 55 and the general points about writing extended answers on page 86.

★ scientists have collected a lot of evidence for this idea

★ the evidence fits/supports the theory

★ scientists are still discovering new evidence that supports the theory

★★ similar species of animals with slight differences are found in environments with slight differences

★★★ if ground becomes drier and harder, animals with smaller feet can move faster across it. Over time, the animals that could move faster to escape predators survived and passed this characteristic on to their offspring. Horse fossils show a gradual change from splayed toes (to provide support on wet ground) to small hooves.

★★★ if vegetation changes from softer leaves to harder grasses, animals with larger and harder teeth can feed better. Over time, the animals that could better chew grass survived and passed this characteristic on to their offspring. Horse fossil teeth show a gradual increase in size and hardness.

★★★ if grasslands replace forests, taller animals get a better view of predators. Over time, the animals that could better see predators survived and passed this characteristic on to their offspring. Horse fossils show a gradual increase in size.

B4 FISHY TREATMENTS

1 A reduction in the symptoms of psoriasis depends on fish/*Garra rufa* eating dead skin. *Or you could have written* 'Garra rufa *can help psoriasis'. You don't have to use 'depends on' but it often helps you to write a hypothesis correctly.*

2 People swimming in the Kangal hot spring in Turkey.

3 • dependent (output) variable: amount of psoriasis or reduction in pain/itching/skin scaliness
• independent (input) variable: use of skin-eating fish
• control variables: type of fish, length of time of treatment. *You'll notice that there are some other control variables that don't seem to have been controlled, such as the number of fish and the age of the patients. Since there was no control (with no use of the skin-eating fish), it's quite difficult to see what the independent variable is.*

4 3

5 Two from:
• to work out the range of measurements needed
• to work out the interval needed between the measurements
• to work out how many measurements to take
• to make sure a method works/to make sure you have the correct apparatus
• to make sure that you have accurate enough measuring apparatus
• to make sure that your investigation is safe.

6 a Each calculation needs to be done this way:

$$\frac{\text{Number who answered a particular way}}{\text{Total number who answered that question}} \times 100 =$$

e.g. 46 people answered 'extremely' to the question 'Was there a reduction in itching?' and 63 people in total answered this question. So,

$(46/63) \times 100 = 73\%$

The table shows the rest of the answers:

Was there a reduction in ...	Number who answered	Answered: 'Extremely'	Answered: 'Considerably'	Answered: 'A little'	Answered: 'Not at all'
... itching?	63	73%	24%	3%	0
... pain?	54	89%	11%	0	0
... skin scaliness?	66	85%	13%	1%	0

b Your bar chart should have the bars grouped together for each of the three questions.

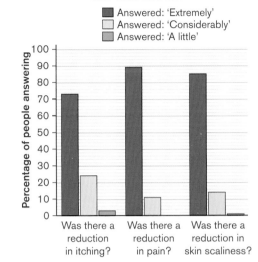

Answers to questions about treatment with *Garra rufa* fish

■ Answered: 'Extremely'
☐ Answered: 'Considerably'
▨ Answered: 'A little'

Make sure you have:
- *a title*
- *the dependent variable on the y-axis*
- *the independent variable on the x-axis (in this case the independent variable is the use of fish and each of the bars shows what happened when the fish were used)*
- *a good scale for the y-axis so that the plotted points are well spread*
- *a y-axis scale that has even divisions*
- *a y-axis scale that is numbered*
- *a label for the x-axis*
- *bars grouped together in threes on the x-axis*
- *a label for each group of three bars on the x-axis*
- *gaps between the x-axis bar groups*
- *all bars accurately plotted*
- *all bars neatly drawn with a ruler*
- *a key to show which bar is which in each group.*

The angles you need for the pie charts (shown on the next page) are calculated as shown in the following tables.

Was there a reduction in ... itching?	Answered: 'Extremely'	Answered: 'Considerably'	Answered: 'A little'	Total
percentage	73	24	3	100
angle calculation 360/100 = 3.6	73 × 3.6	24 × 3.6	3 × 3.6	
angle	263°	86°	11°	360°

Was there a reduction in ... pain?	Answered: 'Extremely'	Answered: 'Considerably'	Answered: 'A little'	Total
percentage	89	11	0	100
angle calculation 360/100 = 3.6	89 × 3.6	11 × 3.6	0	
angle	320°	40°		360°

Was there a reduction in ... skin scaliness?	Answered: 'Extremely'	Answered: 'Considerably'	Answered: 'A little'	Total
percentage	85	14	1	100
angle calculation 360/100 = 3.6	85 × 3.6	14 × 3.6	1 × 3.6	
angle	306°	50°	4°	360°

Make sure each pie chart (see next page) has:
- *the correct angles*
- *a title*
- *all the categories neatly labelled*
- *lines drawn with a ruler*
- *circle drawn with a compass or similar.*

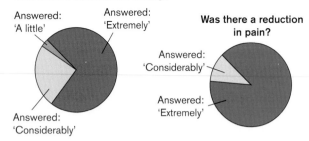

Was there a reduction in itching?

Answered: 'A little'
Answered: 'Extremely'
Answered: 'Considerably'

Was there a reduction in pain?

Answered: 'Considerably'
Answered: 'Extremely'

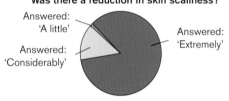

Was there a reduction in skin scaliness?

Answered: 'A little'
Answered: 'Considerably'
Answered: 'Extremely'

  c The bar chart allows a better comparison to be made between the questions.

7 Those reading the paper can be more certain that the results and conclusions are correct because they have been checked by other scientists.

8 **[WEA]** *This is an 'evaluate' question. To answer questions like this you need to use the information that you are given and pick out the evidence for and against the claim. Try to develop an argument. You should finish with a conclusion to say whether you think the claim is justified or not.*

Check your spelling, grammar and punctuation and try to include scientific words, such as 'control', 'data' and 'variable'.

You should try to include some of the points in the list below. You don't need to make all the points but you should aim to make points with the more stars. Also look at the grade booster box on page 56 and the general points about writing extended answers on page 86.

★ weaknesses in the study include: number of fish not controlled, age of patients not controlled, not very many people in the study, there is no control group

★ the claim on the website is invalid

★ the numbers in the table show that people thought that the fish were helping

★★ if not all the control variables are controlled you cannot be sure that you are only measuring the effects of the fish (unless you have a control group)

★★ there is no control group and so the study assumes that without the fish the people's psoriasis would not improve

★★ you can be more sure of the data from bigger studies

★★★ the claim on the website is invalid because the study looked at a combination of fish and sunbed, but the claim is for fish-only treatment

★★★ without a control group you can't be sure that the fish have any effect (it might be time that makes the psoriasis better, the feeling of being cared for, or the sunbed)

★★★ if the number of fish are not controlled then you can't be sure whether you need a certain number of fish to get a benefit

★★★ if ages are not controlled the effects on older and younger people may be different, and so you cannot claim that it will be a beneficial treatment for all ages

★★★ anomalous results are more likely to affect the conclusions in a small study because these will have a greater effect when included in calculations, such as calculating means

B5 BIG BRAINS

1 a cm^3

  b 'centimetres cubed' or 'centimetres × centimetres × centimetres'

2 $650\,cm^3$ and $804\,cm^3$. *This question tests your ability to read off scales but don't forget the units. Notice that the x-axis starts 1 800 000 years ago and you get closer to today as you go left along the axis.*

3 a $1000\,cm^3 - 1300\,cm^3$

  b (1000 + 1030 + 1100 + 1300)/4 = 4430/4 = 1107.5 = 1110 (3 sf) or 1100 (2 sf) *You can't really give your answer to more than 3 significant figures because the figures you've used in the calculation are all given to 3 significant figures or less.* 🔺 *Even if you get this answer wrong, you'd still get some credit for the working … and don't forget the units!*

  c Plot the means on a scatter graph and use error bars to show the ranges.

4 a quantitative **b** secondary

5 The more recent the skull the bigger its inside volume (a negative correlation between age and volume).

6 1.8×10^6

7 Draw a line of best fit; choose two points on the line that are easy to read; use these two points to calculate a gradient, (change in volume/change in number of years); this will give you a rate of change in cm^3/year.

8 a People who have only half the amount of brain that is normal are still able to perform the complex tasks that anyone else can. *The clue here is to remember to look at all the information on the page, including Figure B.*

b *This is quite tricky! An assumption is something that you think is accepted as correct and so you don't try to show that it's correct.* Three possible assumptions (of which you need two) are:

- the bigger a skull is the more brain it will contain
- bigger brains are needed for more complex behaviour
- the skulls that have been measured are actually from the ancestors of humans.

9 Because the measurements have been done by many different scientists who have probably used slightly different ways of measuring the volumes.

10 WEA *This question asks you to justify your choice. This means that you have to weigh up the two choices, come to a decision and use evidence to back up your choice.*

Check your spelling, grammar and punctuation and try to include scientific words, such as 'evidence', 'data' and 'theory'.

You should try to include some of the points in the list below. You don't need to make all the points but you should aim to make points with the more stars. Also see the grade booster box on page 58 and the general points about writing extended answers on page 86.

★ choose one of the ideas
★ state that the data that shows this is in the scatter graph
★★ describe the shape of the data points in the graph to support a choice
★★ make a reference to how good the data is (which may include the fact that a better judgment could be made if there were more data)
★★★ suggest how one or more lines of best fit could be drawn through the data

★★★ explain how a line of best fit should be drawn and use this to justify your choice of line or lines of best fit to support your argument
★★★ describe the validity of the data (which may include mentioning one or more of the assumptions from question 8b).

B6 PLANTS IN SPACE

1 *In a question like this, there would usually be 1 mark for the explanation and 1 mark for giving an example. But the question asks you to 'illustrate', which means that you can't just state the example, you have to use it as part of your explanation of why a model is useful.*

A model makes a complex idea easier to understand. The chemical equation for photosynthesis shows the reactants and the products without all the very complicated processing that takes place to convert one to the other.

2 Glucose molecules contain 6 carbon atoms, 12 hydrogen atoms and 6 oxygen atoms.

3 Test 3 for green light. It is an anomalous result or outlier and is so way off that it's unlikely to be accurate. If you included it in mean calculations it would alter any conclusions that might be drawn from the results. *Remember that the question asked you to 'explain' so don't simply 'state'.*

4 Red. *They are the readings that are the most closely grouped.*

5 a $58\,mm^3$

b The means are: white $67\,mm^3$/min, blue $58\,mm^3$/min, green $20\,mm^3$/min, yellow $23\,mm^3$/min, red $51\,mm^3$/min. *All these are given to 2 significant figures because that's the number of significant figures we have in the data. And the figure for green misses out the anomalous result/outlier because this could have been caused by writing down the figure wrong (e.g. adding a 0 by mistake, making it 180 and not 18).*

6 mm^3/min. *This is probably the most useful measurement with the figures that you have got although you could divide everything by 60 to get mm^3/s but that's quite a lot of extra work for no real benefit.*

7 You need to draw a bar chart because the independent variable is categoric and the dependent variable is continuous.

You might have included error bars, which are always a nice touch but you won't lose credit for not including them.

Make sure you have:
- a title
- the dependent variable on the y-axis
- the independent variable on the x-axis
- a good scale for the y-axis so that the plotted points are well spread
- a y-axis scale that has even divisions
- a y-axis scale that is numbered
- a label for the y-axis (with units)
- a good scale for the x-axis so that the plotted points are well spread
- a label for the x-axis
- gaps between the x-axis bars
- the categories correctly labelled on the x-axis
- all bars accurately plotted
- all bars neatly drawn with a ruler (they needn't be coloured).

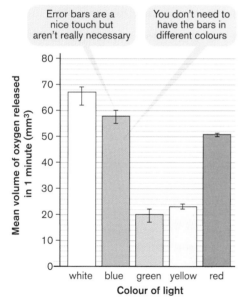

Oxygen released by pond weed in different colours of light

Error bars are a nice touch but aren't really necessary

You don't need to have the bars in different colours

8 *In a question like this, there would usually be 1 mark for the explanation and 1 mark for giving an example. But the question asks you to 'illustrate', which means that you can't just state the example, you have to use it as part of your explanation of why a control is useful.*

The results from the control form a baseline to compare with the other results. In this experiment, the white light is the control. We know that plants grow well in white light and so we can compare the other colours of light against it.

9 WEA *When answering questions like this, think about financial costs, effects on people, effects on the environment and ethics. The question asks you to discuss, so you should structure your answer carefully and not just give a list of points.*

Check your spelling, grammar and punctuation and try to include scientific words, such as 'photosynthesis'.

You should try to include some of the points in the list below. You don't need to make all the points but you should aim to make points with the more stars. Also look at the grade booster box on page 59 and the general points about writing extended answers on page 86.

★ indicate which colour gave the highest numbers/most oxygen

★ glucose is made during photosynthesis and the more of this that can be made in a short period of time the better

★★ the chemical equation tells us that the more oxygen being released by a plant, the faster it is photosynthesising

★★ other factors need to be taken into account such as the cost of the LEDs, the amount of heat they release

★★ in making a decision you need to balance the benefits and the drawbacks

★★★ you need evidence about the costs (e.g. one LED might allow a slightly greater rate of photosynthesis than another but cost a lot more, or have a shorter life-span)

★★★ you need evidence about the other effects of the LED (e.g. one LED might allow a slightly greater rate of photosynthesis than another but produce too much heat, or use much more electricity)

★★★ you need evidence about environmental effects (e.g. will the LED's manufacture damage the environment?)

★★★ you need evidence about safety (e.g. will the LED be safe to use?)

B7 HEALTH SCARE

1 The graphs are a line graph combined with a bar chart of grouped continuous data. *All the data for measles and MMR are combined into 'year groups'. Since it is continuous data you could plot both sets of data as line graphs. Since the data is grouped into years you could also plot both sets*

of data as histograms. Combining a histogram with a line graph can be useful because it makes the two sets of data look different on the graph.

2 A paper that has been checked by an expert in the same subject before being published.

3 a Money. *From the information on the page this is the most likely reason.*

  b There may be bias in his data caused by too few samples/children.

4 a As the numbers of children who are not immunised falls, the cases of measles rise.

  b Yes. There is a good explanation for the correlation – people who are immunised do not get the disease, therefore if fewer people are immunised then it follows that more will get the disease.

5 You could have made any one of these points:
  - the study looked at only 12 children (the sample size was too small)
  - there was no control group
  - it was based on what parents had told him (rather than on experiments)
  - parents of 4 out of the 12 (33%) did not link autism to the MMR vaccination.

6 You could have made any one of these points:
  - he should not have taken blood at a birthday party by paying children
  - you need permission from parents to collect a child's blood and it should be done in a hospital
  - he received money from the parents' solicitors but didn't mention that when he published his results.

7 Other scientists cannot repeat his findings.

8 **WEA** *If a question asks you to explain something then take great care that you <u>explain how</u> something is happening, and you don't just <u>describe what</u> happens.*

  Check your spelling, grammar and punctuation and try to include scientific words, such as 'vaccination' or 'immunisation'

  You should try to include some of the points in the following list. You don't need to make all the points but you should aim to make points with the more stars. Also look at the grade booster box on page 60 and the general points about writing extended answers on page 86.

★ reports in the media can change people's minds about scientific things

★ many people rely on the media to make sense of science

★ a claim may be widely reported in the media and therefore many people find out about it and start to discuss the subject

★★ a claim might cause panic for no need

★★ Andrew Wakefield's claim that MMR could cause autism meant that many people stopped having their children vaccinated

★★★ Andrew Wakefield's claim that MMR vaccinations could cause autism led to more children getting measles

★★★ some media stories about Andrew Wakefield's claim did not question his paper or his claims with the result that there was an over-reaction by people

B8 ESTIMATING POPULATION SIZE

1 quadrat

2 a a rough calculation

  b two rectangles (100 × 110) + (20 × 75) plus one triangle (0.5 × 20 × 25) = 11 000 + 1500 + 250 = 12 750 m^2

  c Area of quadrat is 0.5 × 0.5 = 0.25 m^2. Total sampled area = 0.25 × 15 = 3.75 m^2. Daisies = 57/3.75 = 15.2 plants/m^2. Total population = 15.2 × 12 750 = 193 800 plants.

  🔺 A better answer would use 2 or 3 significant figures.

  d 21/3.75 = 5.6 plants/m^2. Total population = 5.6 × 12 750 = 71 400 plants.

3 a Two of: the size of the playing field; the ease of counting the plants inside a quadrat; the amount of time they had; the accuracy of the result that they needed.

  b 3400:1

  c No. The sample area is tiny compared with the total area.

4 a Flipping a coin to see whether to throw the quadrat left or right.

  b So that the results are not affected by bias.

c They may have looked when throwing the quadrat, helping to make sure it landed in a certain place.

d Divide the area into a grid and then generate random numbers to choose grid squares to sample.

5 It is not made using the results that have been presented. A valid conclusion can only be based on the results that have been presented.

6 a One sensible suggestion, for example: making sure no one was in the way when the quadrat was thrown; washing hands after doing the experiment; wearing gloves.

b One sensible suggestion, for example: avoid digging up/picking the plants while counting them.

B9 MORE THAN A TAN

1 Why is the number of people who get malignant melanoma increasing?

2 a The levels of malignant melanoma depend on the number of holidays abroad.

b One of:
- there is not very much data
- you need more than 6 years to be sure of the trend
- the data does not show that the people who go on more holidays abroad are the ones that are getting more malignant melanomas.

3 Compare only the years 2003 to 2005, when the ozone levels are falling (which means that there will be more UV reaching the Earth). *Not showing all the data is a common way of biasing evidence.*

4 a 6.95×10^7. *The number from the graph is 69.5 million, which is 69 500 000.*
🔺 *You'll get credit for showing your working, even if you misread the graph.*

b $(69.5 − 61.4)/(2006 − 2003) = 8.1/3 = 2.7$ million holidays per year (or 2 700 000 holidays per year). *Don't forget to read the scales properly. In this case the y-axis is in 'millions' and it's easy not to spot this. The rate of increase is the slope of a straight line between the point for 2003 and the point for 2006 on the graph. This is worked out by calculating the amount of change of the y-variable and dividing this by the amount of change of the x-variable.*

5 *You'll have needed a ruler for this, to see a line of best through the points.*
- for 2002, any number between 11.4 and 11.8
- for 2009, any number between 16.8 and 17.2

6 Benefit: helps the skin to make vitamin A.

Drawbacks: skin cancer/malignant melanoma/ sunburn.

Risk: risk of malignant melanoma is 1 in 77 for women or 1 in 91 for men. *Remember that the risk should be linked to the drawback.*

7 1.1% *There is a 1 in 91 chance, which can be written as 1/91. You multiply the result by 100 to get the percentage. And since the data you are using are to 2 significant figures your answer should be too.*

8 **WEA!** *Check your spelling, grammar and punctuation and try to include scientific words, such as 'melanoma'.*

You should try to include some of the points in the list below. You don't need to make all the points but you should aim to make points with the more stars. Also look at the grade booster box on page 62 and the general points about writing extended answers on page 86.

★ run a survey where you ask people how long they use a sunbed for on average each month or find secondary data that contains this information

★ include in the survey a question about whether the people have or have not had malignant melanoma or find secondary data that contains this information

★ see if there are more cases of melanoma in people who use sunbeds more

★★ the group of people who have never used sunbeds is the control

★★ a control is when the independent (input) variable is not applied

★★ a control lets you see if changes are due only to the independent (input) variable

★★ divide the data for those who use sunbeds into groups, depending on the number of hours spent on the sunbed

★★★ calculate the probability that a person in a group will get a malignant melanoma

★★★ use a histogram to spot a correlation

★★★ calculate the probability of someone in the control group getting malignant melanoma and see whether the probabilities in the other groups are higher than this

B10 BMI

1 a *BMI is calculated using metres and not centimetres, so you first need to convert 181 cm to 1.81 m.*

BMI = $99/1.81^2$ = $99/(1.81 \times 1.81)$ = 30.2 so Derek is obese.

b BMI = $43/1.36^2$ = $43/(1.36 \times 1.36)$ = 23.2 so (from Figure B) Jane is overweight.

All these answers are given to 3 significant figures because that's the number of significant figures we have in the data. And don't forget that you need to use the table for 1a and Figure B for 1b because of the ages of the people.

2 a 'square metres' or 'metres squared'

b area

3 A rough calculation. *A BMI cannot be any more than an estimate because your mass varies from day to day depending on when you last ate, the time of day, etc.*

4 Any one of: number of clothes children are wearing when having their masses measured; the time of day they are measured; the time of year they are measured; the resolution of the 'scales' used; the accuracy of the scales used.

5 If a large number of measurements is missing it can cause bias. *Think about it this way. If the reason most of the people who didn't want to be measured was because they were embarrassed about their masses, then there will be a large number of overweight people whose masses are not used in the graph. This will mean that there is bias towards lighter people.*

6 The more wealthy people are, the less likely they are to be obese. *If you plotted a graph of 'wealth' on the x-axis against 'obesity' on the y-axis, the line of best fit would slope downwards.*

7 **WEA** *The question needs you to make two suggestions: one for why the company advertised it in that way and the other for why it was banned.*

Check your spelling, grammar and punctuation and try to include scientific words, such as 'nutritious'.

You should try to include some of the points in the list below. You don't need to make all the points but you should aim to make points with the more stars. Also look at the grade booster box on page 63 and the general points about writing extended answers on page 86.

★ the company wanted to encourage people to buy the drink

★ the drink contains vitamins and your body needs vitamins

★ the drink contains sugar, which your body needs

★★ the word 'nutritious' makes the drink sound healthy

★★ people do not expect something that is healthy to contain so much sugar

★★ consuming too much sugar over a long period can cause obesity/diabetes

★★★ there is no evidence to support the claim that the product is actually good for you

★★★ drinking so much sugar in one serving is not good for you because it means you are likely to consume too much sugar in a day

★★★ anything that is nutritious contains sources of 'energy supply' substances and substances needed for health, which this product contains

C1 MEMORY OF WATER

1 a To make complex ideas easier to understand.

b Good: it shows the number of atoms in the water molecule, it shows how the atoms in a water molecule are arranged, it shows which atoms are found in a water molecule.
Poor: the atoms don't have those colours, the atoms are not joined with 'sticks'.

2 2:1 *Make sure you write it this way round.*

3 Mix one volume of the original chemical with nine volumes of water to make a 10 × dilution. Take one volume of the 10 × dilution and mix it with nine volumes of water to make a 100 × dilution. Take one volume of the 100 × dilution and mix it with nine volumes of water to make a 1000 × dilution (1×10^3). Take one volume of the 1000 × dilution and mix it with nine volumes of water to make a 10 000 × dilution (1×10^4).

OR Take one volume of the original and mix it with 9999 volumes of water.

OR Take one volume of the original and mix it with 99 volumes of water to form a 100 × dilution and then take 1 volume of the 100

× dilution and mix that with 99 volumes of water. *There are lots of ways of doing this. Just make sure that you understand that a 1 × 10⁴ dilution means 1 part in 10 000, which is the same as saying 1 volume of the original mixed with 9999 parts of water forming a total of 10 000.*

4 a Any answer between 1:1.8 and 1:1.9 for the ratio of the range of the graph on the left to that of the one on the right. You need to divide the range of the second graph (which is 73 − 0) by the range of the first graph (40 − 0), i.e. 73/40 = 1.8. So the ratio tells you that the range of the second graph is 1.8 times bigger than the range of the first.

b They show that when the person counting the cells does not know which dilution the cells have been in, the experiment does not work as expected. For instance, in the left-hand graph there is a peek at '6'. On the right-hand graph this is a trough.

5 Two from: type of cells; type of water used; chemical used; containers in which chemicals/cells were mixed; method of counting cells without granules.

6 Neither he nor the reviewers could find any fault with the method and the results as they were set out in the paper. *This doesn't mean that there wasn't something wrong, just that nobody could tell what was wrong.*

7 [WEA] *The question asks you to use information from the page. You may not get any marks if you don't do this.*

Check your spelling, grammar and punctuation and try to include scientific words, such as 'error' and 'bias'.

You should try to include some of the points in the list below. You don't need to make all the points but you should aim to make points with the more stars. Also look at the grade booster box on page 64 and the general points about writing extended answers on page 86.

★ the person counting the cells made mistakes
★ this is called human error
★★ the results in the paper show systematic error
★★ the results from the two graphs in Figure B show random error
★★ the results shown in the paper show bias
★★★ the person counting the cells was

convinced that certain dilutions had effects on the cells and so subconsciously counted more cells without granules in those dilutions compared to dilutions that she didn't think had an effect

★★★ by leaving out the results from experiments that don't work, you create bias shifting all the results in a certain direction

★★★ the peaks are bigger than they should be, since they are calculated from means of measurements but without the results from experiments that didn't work

C2 THE PERIODIC TABLE

1 He predicted:

- elements that had not yet been discovered
- that some of the 'atomic weights' were wrong.

You could have used the 'if … then …' phrase in your answer:

- *If there are gaps when elements with similar properties are put together, then there must be undiscovered elements to fill those gaps.*
- *If two elements are in order of their 'atomic weight' but don't match the properties of the element above them, then their 'atomic weights' are wrong.*

2 It did not explain all the data (he left out half of the known elements).

3 a Ga (gallium). *You'll have needed to look this up in the periodic table on page 126.*

b Mendeleev had predicted that this element would be found and so this supports his theory/idea.

c The graph shows that elements follow a pattern when put in 'atomic weight' order. This is the same pattern that Mendeleev found. *Or you could have mentioned that Lothar Meyer reached the same conclusion as Mendeleev but was working independently.*

4 a line graph

b The mass could be measured in grams or milligrams and the volume in cm^3 or mm^3.

5 a secondary

b One of: you have no control over the quality of the data; it doesn't cost you much money or time to obtain it; it's easy to obtain.

6 One of: they are quick to write and so often

clearer; they aren't affected by different names/spellings in different countries; all scientists across the world understand instantly what they mean.

7 You should have 602 with 21 zeros after it.

8 **WEA** *This question has two parts. The first part wants you to describe a process and so you will need to write what happens in a logical way. The other part wants you to talk about the benefits, so you need to refer to your description of the process and pick out the parts that are beneficial.*

Check your spelling, grammar and punctuation and try to include scientific words, such as 'journal' and 'paper'.

You should try to include some of the points in the list below. You don't need to make all the points but you should aim to make points with the more stars. Also look at the grade booster box on page 65 and the general points about writing extended answers on page 86.

★ a scientist's paper is checked by other scientists

★ this means that the reader can be more sure that the paper contains good science

★★ the paper is sent to a journal

★★ the editor of the journal sends it out to other experts in the same field

★★ comments from experts mean that the paper is published, amended and then published, or not published

★★ a reader does not have to carefully analyse the paper for mistakes since this has been done by experts

★★★ the investigations are evaluated to ensure that the results are valid

★★★ the conclusions are checked to make sure they can be drawn from the results

★★★ the reviewer looks out for bias

C3 TESTING MATERIALS

1 a 92.5%

b *You need to have drawn a pie chart for this data because you are showing the different proportions that contribute to a whole.*

The angles you need for the pie chart are calculated as follows.

Element	Al	Si	Mg	Total

percentage	92.5	7.1	0.4	100
angle calculation 360/100 = 3.6	92.5 × 3.6	7.1 × 3.6	0.4 × 3.6	
angle	333°	26°	1°	360°

Make sure each pie chart has:
- *the correct angles*
- *a title*
- *all the categories neatly labelled*
- *lines drawn with a ruler*
- *circle drawn with a compass or similar.*

Percentages of elements in an aluminium alloy

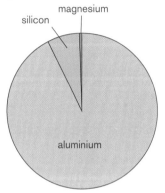

2 Two from: keep the testing machine the same; use the same shape of alloy bar each time; increase the force on the bar at the same rate for each bar.

3 a 40 kN (or 40 000 N). *You'll have needed to extend the line of best fit to the y-axis.*

b 69.4 kN and 55.5 kN because they do not sit close to the line of best fit like the other measurements.

4 56 000 N, 69 000 N, 71 000 N, 80 000 N, 97 000 N
Don't forget that the question wanted the answer in newtons and not kilonewtons.

5 a Volume = 1 × 2 × 2 = 4 cm³
Density = mass/volume = 32/4 = 8 g/cm³

b Density = mass/volume and so mass = density × volume = 2.7 × 4 = 10.8 g.
OR Use ratios: mass = (2.7/8) × 32 = 10.8 g.
Don't forget the unit in your answer.

6 The benefit is that they are lighter than normal/steel wheels. The drawback is that they are more expensive. *You might be able to think up other benefits and drawbacks but these are the ones given on the page.*

7 **WEA** *The question asks you to evaluate a*

number of things – the conclusions. So you need to compare the different conclusions, pointing out their good and bad points, before choosing one conclusion and saying why you have chosen that one.

Check your spelling, grammar and punctuation and try to include scientific words, such as 'valid'.

You should try to include some of the points in the list below. You don't need to make all the points but you should aim to make points with the more stars. Also look at the grade booster box on page 66 and the general points about writing extended answers on page 86.

★ conclusion ii selected

★ points out that two conclusions are for a crane hook and one is for a bridge support

★ points out that two conclusions are for 7.5% manganese

★ points out that 7.5% manganese is the strongest alloy that was tested

★★ points out that the test is for stretching the bar and not squashing it

★★ points out that 7.5% manganese was the highest proportion of manganese tested

★★ points out that conclusions i and iii are invalid

★★★ dismisses conclusion i because the strength of the bar when squashed is not tested

★★★ dismisses conclusion iii because the conclusion is not based on the test results. *It's impossible to say that the bar keeps on getting stronger the more manganese you add to it. From the results that you have, you could predict that this will be the case but you would need to test it before you can draw that conclusion.*

C4 CHEMISTRY AND WAR

1 a MPa, megapascals

b 1 megapascal is 1 000 000 pascals, and you could explain that 1 pascal is the same as the force of 1 N on 1 m².

2 a The higher the temperature the less ammonia formed.

b negative

3 *Your table should be in order of pressure or of percentage of ammonia formed.*

Pressure (MPa)	Percentage of ammonia formed (%)
20	19
30	27
41	32
51	36.5

4 a It's poisonous or high pressures might cause an explosion.

b Make sure that the ammonia is always carefully contained and cannot escape or make sure the reaction vessel is very strong. *Make sure you understand the difference between a hazard and a risk.*

5 a percentage ammonia formed, pressure, temperature

b percentage ammonia: dependent, quantitative, continuous
pressure: independent, quantitative, continuous
temperature: independent, quantitative, continuous

6 You should try to include some of the following points:

- a balance between speed, expense and percentage of ammonia formed must be reached
- although the temperature could be lower to increase the percentage of ammonia, this makes the reaction too slow
- although the pressure could be higher, this requires more expensive equipment and the increase in percentage isn't worth it

7 **WEA** *The question asks you to write an argument. So you first need to plan out what points you are going to make and what points others might use to argue against you.*

Check your spelling, grammar and punctuation and try to include scientific words, such as 'ammonia' and 'Haber process'.

You should try to include some of the points in the following list. You don't need to make all the points but you should aim to make points with the more stars. Also look at the grade booster box on page 67 and the general points about writing extended answers on page 86.

★ states that you are in favour of the factory

★ gives a reason for being in favour of the factory (e.g. it will improve local facilities, it will provide jobs)

★★ explains one or more reasons against the factory (e.g. noise will be caused by transport to and from the factory, building work will destroy habitats for animals and plants, ammonia is poisonous and might leak)

★★ explains one or more reasons in favour of the factory (e.g. it will improve local facilities because that factory will need to build new roads/houses/facilities for the workers, the factory will need people to run it and so it will provide jobs)

★★★ writes an argument with the same structure as shown in S60

★★★ uses personal knowledge to back up reasons in favour of the factory (e.g. it will provide jobs because whenever something new opens they need people to operate the machinery and this factory is new to the area – it's not replacing something that was already there)

★★★ writes a clear response to one counterargument (e.g. some people might say that ammonia is made to make dangerous explosives, but the explosives will not be made at the factory and most ammonia is used to make fertilisers which are important)

C5 RIVER QUALITY

1 Qualitative: grades; quantitative: either the percentage oxygen or the ammonia concentration.

2 A molecule (or unit) of ammonia contains 1 nitrogen atom and 3 hydrogen atoms.

3 less than

4 mg/dm^3 is a compound unit. *Remember that a compound unit or measure is nothing to do with chemical 'compounds' – it is a measuring unit made by combining other units.*

5 A total of 36 readings gives a good number of samples which means that the results are less likely to be biased by anomalous results.

6 Chemical fertilisers increase ammonia levels in rivers and so can harm water creatures. Also chemical fertilisers cause algae growth, which causes problems when the algae die because

the bacteria that break them down use up a lot of the oxygen, meaning there is much less for other water creatures.

7 a No, because they vary a lot (for example, the readings for dissolved oxygen vary from 75 or below to 94 or above).

b **WEA** *This question asks you to state something so make sure that you give short and accurate meanings before grading the river. It then asks you to explain how you do something, so make sure that you explain all the steps in how you would use these figures to work out a grade for the river.*

Check your spelling, grammar and punctuation and try to include scientific words, such as 'percentile'.

You should try to include some of the points in the list below. You don't need to make all the points but you should aim to make points with the more stars. Also look at the grade booster box on page 68 and the general points about writing extended answers on page 86.

★ a grade of A or B is given (A isn't correct but very nearly is)

★ the ammonia readings for the river are all under the $0.25\,mg/dm^3$ limit for a grade A

★★ the dissolved oxygen is 79% at the 10th percentile but needs to be 80% for a grade A

★★ the dissolved oxygen percentage scores the river a grade B but the ammonia concentration scores a grade A. The overall grade is the lower of the two (i.e. B).

★★★ a percentile is the cut-off point for a certain percentage of all the readings (or is the measurement at or below which a certain percentage of readings fall)

★★★ a 10th percentile is the reading at the 10% level of all the measurements, cutting off the bottom 10% of all readings OR a 90th percentile is cutting off the bottom 90% of all readings

★★★ for a grade A for dissolved oxygen, only 10% of readings can be below 80%

★★★ for a grade A for ammonia, 90% of readings must be below $0.25\,mg/dm^3$

C6 RATES OF REACTION

1 calcium chloride + hydrochloric acid → calcium chloride + water + carbon dioxide

2 measuring cylinder

3 a surface area/size of the chips

b mass

c two of: concentration of acid, amount of acid, temperature, (mass of calcium carbonate)

4 3.1/0.52 = 6.0 (to 2 significant figures)

5 a (49.9 − 52.7) / 52.7 = − 0.0531 = − 5.31% (or 5.3%)

b (50.4 − 53.9) / 53.9 = 0.0649 = 6.49% (or 6.5%)

c Not very accurate because you would expect the percentage change for the same reaction to be the same.

6 a The greater the surface area (of the calcium carbonate/reactants) the faster the rate of reaction.

b Smaller chips would make the reaction go faster.

c Yes, although the data collected is not very good.

7 Making repeat readings, ensuring that the same masses of the reactants are used each time.

8 Flask A: (52.0 − 52.7)/10 = −0.07 g/s
Flask B: (52.4 − 53.9)/10 = −0.15 g/s

9 a An appropriate graph with curved lines of best fit through the points to show both sets of results.

b Tangent drawn and used to calculate a gradient; expected gradient is about −0.1 g/s (0.1 g/s mass lost).

C7 EARTH WARMING

1 You need to draw a bar chart with grouped continuous data. And note that not all of the groups have the same range, so not all the bars will have the same width. ▲ It would therefore be better to plot a histogram, with frequency density on the y-axis (number of times the barrier was shut/year).

Make sure you have:

- *a title*
- *the dependent variable on the y-axis*
- *the independent variable on the x-axis*
- *a good scale for the y-axis so that the plotted points are well spread*
- *a y-axis scale that has even divisions*
- *a y-axis scale that is numbered*
- *a label for the y-axis*
- *a good scale for the x-axis so that the bars are wide*
- *a label for the x-axis*
- *no gaps between the x-axis bars*
- *an x-axis scale that has even divisions*
- *the categories correctly labelled on the x-axis (you may have drawn a scale as shown in the diagram here, or labelled these groups on the x-axis as '1980–1989', '1990–1999', '2000–2009', '2010–2015')*
- *all bars accurately plotted*
- *all bars neatly drawn with a ruler.*

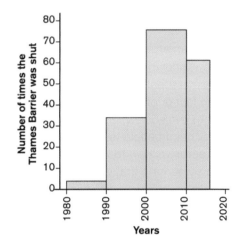

2 A bar chart has been drawn for the left-hand chart because one variable is categoric. A line graph has been drawn for the right-hand graph because both variables are continuous.

3 A way of thinking about something using a computer, which makes things easier for us to understand.

4 a It's quicker/easier just to use one set of data.

b One data set is unlikely to reflect the same conditions as elsewhere in the world and so will produce bias in the models.

c Collect CO_2 data from many places around the world, and locate those places at random.

5 The graph is a fairly straight line so pick two points that are easy to read off it to calculate the gradient, e.g.
$$\frac{(80 - 60)}{(95 - 81)} = \frac{20}{14} = 1.4 \, \text{ppm/year}$$

6 a 1 in 1000 is 1/1000 = 0.001. To get a percentage multiply by 100 giving 0.1%.

b The risk is halved. So, 0.1 ÷ 2 = 0.05%.

c 0.2%. *A tricky one!* If the barrier halves the risk but by 2030 the reduced risk is back to today's risk, then the risk has doubled. So if the Thames Barrier had not been built, the risk of London flooding in 2030 is twice what it is today without the Thames Barrier.

7 The cost of building it was very much less than the cost of the city flooding.

8 **WEA** *This question asks you to discuss something so you need to identify the different points made about the issue – both points in favour and points against. Then build up your own argument about the issue to reach a conclusion or decision.*

Check your spelling, grammar and punctuation and try to include scientific words, such as 'risk'.

You should try to include some of the points in the list below. You don't need to make all the points but you should aim to make points with the more stars. Also look at the grade booster box on page 71 and the general points about writing extended answers on page 86.

★ the existing barrier will not work forever

★ the proposed barrier is very expensive

★★ after 2030 London will be at greater risk of flooding

★★ the use of the current barrier is increasing

★★ global CO_2 levels appear to still be rising, which is thought to be causing melting of ice and this causes sea levels to rise

★★ a second barrier on the Thames may not work

★★★ £20 billion to build the new barrier is a quarter of the cost of damage that a major flood would cause in London

★★★ none of the models predict that CO_2 levels are going to do anything but rise, so sea levels will continue to rise and so flooding will become more and more likely

★★★ sea levels could be rising due to something else (e.g. London sinking) so a barrier may not be the most appropriate or the only solution that is needed

C8 AIRBAGS

1 0.03 seconds. *The question wants your answer in seconds and not milliseconds.*

2 … decompose/break apart/go to give 2 atoms/units of sodium and 2 molecules/units of nitrogen. *To be technical, sodium azide is an ionic solid and so you can't refer to it as 'molecules'. This is why the first part of the answer (given in the question) contained the word 'units'. If you've learnt about moles you could use 'moles' instead of 'units'. Sodium exists as atoms. Nitrogen exists as molecules of two nitrogen atoms bonded together.*

3 a Three from: to work out the range of measurements to take; to work out the intervals needed between each measurement; to work out how many measurements to take; to work out how many repeat readings to take; to make sure that you don't make unnecessary readings in the full investigation; to make sure that your method works; to make sure that you have the correct apparatus; to make sure you have accurate enough measuring apparatus; to make sure the method is safe.

b Mass of gas, temperature. *Look at the graph title and the caption.*

4 a The reading at 40 dm³. This reading is anomalous/doesn't fit the pattern.

b Volume is inversely proportional to pressure (or vice versa), for a fixed mass. $P \propto 1/V$ or $V \propto 1/P$.

c Plot pressure against $1/V$ (or volume against $1/P$).

5 The volume of a cylinder is $\pi \times r^2 \times h$. The height is 26 cm and the radius is 58/2 = 29 cm. So the volume = 3.142 × (29 × 29) × 26 = 68 700 cm³, which is about 69 dm³. From the graph, an airbag of 69 dm³ would give a pressure of about 42 kPa. The bag needs to be bigger.

6 Two from: look at the evidence from countries where airbags are compulsory to see if there are problems; look at the figures to see if airbags save lives/prevent injuries; check that airbags don't make cars too expensive; look at the effect of airbags on the environment; look at the effects of airbags going off by mistake.

7 **WEA** *This question asks you to discuss something, which means that you need to build up an argument about whether it is a good idea or not to fit airbags in cars. You will need to think about the benefits, drawbacks and risks and then put your ideas down in the form of an argument (see S60).*

Check your spelling, grammar and punctuation and try to include scientific words, such as 'sodium azide'.

You should try to include some of the points in the list below. You don't need to make all the points but you should aim to make points with the more stars. Also look at the grade booster box on page 72 and the general points about writing extended answers on page 86.

★ identify one benefit of airbags (e.g. an airbag helps to protect people in cars from injury in a car crash, helps to reduce the costs to the NHS)

★ identify one drawback of airbags (e.g. contains a poisonous chemical, may go off accidentally, makes a car cost more)

★ identify one way in which risks are reduced (e.g. putting sodium azide in a strong container, setting the correct volume of the airbag)

★ state that you are in favour of or against airbags, with a reason

★★ explain two or more benefits of airbags (e.g. airbags help protect people in cars from injury in a car crash by inflating very quickly to stop a person hitting the dashboard/steering wheel, help to reduce the costs to the NHS through making injuries less serious)

★★ explain two or more drawbacks of airbags (e.g. contains a poisonous chemical that can stop people breathing, may go off accidentally leading to an accident because the driver can't see or gets a fright, makes a car cost more due to the materials used to make airbags and additional time required to fit the airbags)

★★ explain two or more ways in which risks are reduced (e.g. fitting airbags reduces the risk of injury in a car crash, putting sodium azide in a strong container reduces the risk of it escaping and so coming into contact with someone, setting the correct volume of the airbag to ensure correct pressure so that the airbag works without injuring someone)

★★★ the risk of kidney injury in a car crash is 139/2864 = 0.049, or 4.9%

★★★ the risk of kidney injury with an airbag is 43/2864 = 0.015, or 1.5%

★★★ the risk of kidney injury with an airbag is reduced by 1.5/4.9 = 0.31, or 31%

★★★ an argument written with the same structure as shown in S60

★★★ a clear response given to one counterargument

C9 AVOGADRO'S BIG IDEA

1 3:1

2 a e.g. g/cm^3, g/dm^3, kg/m^3. *You can have a variety of units here, as long as it is a compound unit consisting of a mass unit divided by a volume unit.*

  b 1.10, 0.0732

  c 1.10/0.0732 = 15.0

  d Today the relative atomic mass of oxygen is one unit greater (16). Or, today the relative atomic mass oxygen is 7% bigger. *You can't just write '16' as your answer because the question has asked you to compare two figures. To calculate a percentage increase, work out the size of the increase and divide it by the original amount and then multiply by 100. So (16–15)/ 15 × 100 = 7%.*

3 Your model should look like Figure A but show: i) hydrogen particles composed of two atoms of hydrogen stuck together, ii) oxygen particles composed of two atoms of oxygen stuck together, iii) two volumes of water.

4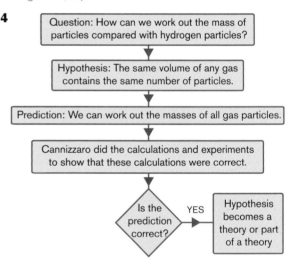

5 Avogadro wasn't very well known and Berzelius was very influential, so people believed him and not Avogadro. Or, he didn't do enough experiments to support his hypothesis.

6 You need to have drawn a scatter graph here to compare the two variables and drawn a line of best fit. From the line of best fit you can calculate its gradient, which will be the volume of gas per mole. See the next page.

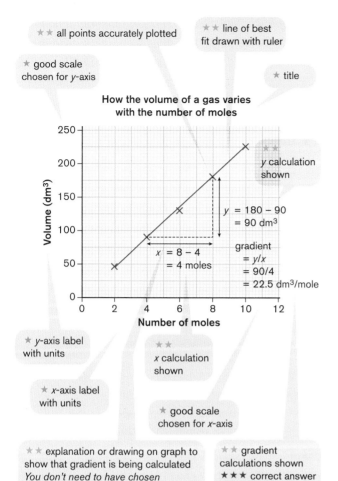

How the volume of a gas varies with the number of moles

$y = 180 - 90$
$= 90 \text{ dm}^3$

$x = 8 - 4$
$= 4 \text{ moles}$

gradient
$= y/x$
$= 90/4$
$= 22.5 \text{ dm}^3/\text{mole}$

C10 COLD PACKS

1 0.1 °C

2 Don't touch/wear gloves and be careful not to raise the dust/don't breathe in the dust/wear a face mask when handling.

3 a The reading of 15.2 °C is well away from all the others.

b 7.6, 8.4, 9.2 (ignoring the anomalous result), 10.8, 10.2 °C

c You need to have drawn a scatter graph because you are looking for relationships/ correlations between two quantitative and continuous variables.

You might have tried a line of best fit for the first 4 points, although this isn't really useful here because the last point doesn't fit the pattern and is not an anomalous result because all the readings for 6 g of sodium nitrate were pretty much the same. It's at this point that you have to think about extending the range of the

masses of sodium nitrate used to see what happens to the temperature drop then.

You might have included error bars, which are always a nice touch but you won't lose credit for not including them.

Make sure you have:

- *a title*
- *the dependent variable on the y-axis (temperature change)*
- *the independent variable on the x-axis (mass of sodium nitrate)*
- *a good scale for the y-axis so that the plotted points are well spread*
- *a y-axis scale that has even divisions*
- *a y-axis scale that is numbered*
- *a label for the y-axis with units*
- *a good scale for the x-axis so that the plotted points are well spread*
- *a label for the x-axis with units*
- *an x-axis scale that has even divisions*
- *an x-axis scale that is numbered*
- *all points are accurately plotted*
- *all points are neatly plotted*
- *(a line of best fit drawn using a ruler, if you've done one).*

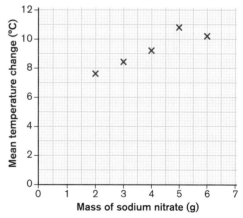

Relationship between the mass of sodium nitrate added and the temperature change

d The more sodium nitrate used the greater the change in temperature.

4 Two from: the apparatus, the starting temperature, the chemicals used, the amount of water added, the amount of ammonium chloride used.

5 More readings need to be taken. The last reading (at 6 g) does not fit the rest of the pattern. This could be anomalous or the change in temperature could have started going down with increasing mass of sodium nitrate. You cannot tell.

6 The cooling ability of a cold pack depends on the chemicals inside it. *That's the hypothesis that the investigation was intended to test. But, if you look at the table, that's not what the investigation actually tested.*

7 [WEA] *This question asks you to evaluate a single thing. So you need to say how good or poor the investigation is based on the list of things to check given in S54.*

Check your spelling, grammar and punctuation and try to include scientific words, such as the correct terms for the units.

You should try to include some of the points in the list below. You don't need to make all the points but you should aim to make points with the more stars. Also look at the grade booster box on page 74 and the general points about writing extended answers on page 86.

★ the data is good because the measurements have been repeated

★ the data is precise because repeated measurements (apart from one) are quite close

★ the conclusion does not match the hypothesis

★ the heading in the table 'change in temperature' is not good because it doesn't tell you whether the change is up or down

★★ the investigation is not valid

★★ the measuring devices were accurate enough for this investigation

★★ there is bias in all the '1st try' readings, which are all slightly on the high side

★★★ the results were accurate enough to draw a conclusion because the changes in the readings are much greater than the resolution of the measuring devices

★★★ the investigation is not valid because the investigation does not test lots of different chemicals

★★★ the only conclusion that can be drawn from these results is that ammonium chloride is better to use in cold backs than sodium nitrate. *This is a difficult one to get! Have a look at the graph you drew for question 3. If you draw and extend a line of best fit (missing out the last point) so that it meets the y-axis, you will see that with no sodium nitrate the temperature change is about 5 °C. This is caused by 2.5 g of ammonium chloride. 2.5 g of sodium nitrate causes an additional temperature decrease of 2.9 °C (about 7.9 °C in total). So ammonium chloride is more*

effective than sodium nitrate. However, this conclusion doesn't really mean this is a good investigation of the hypothesis – there are too few chemicals tested.

P1 COLD FUSION

1 Benefits: electricity without smoke pollution/ without carbon dioxide; drawbacks: produces radioactive waste.

2 a Based on the information on this page you could have any one of three hypotheses: The ability of deuterium to fuse at room temperature depends on the amount of D_2O/ the volume of palladium/the substance used to absorb it. *You don't have to use 'depends on' but it often helps you to write a hypothesis correctly.*

b Two from:
- if deuterium fuses then heat would be produced
- if deuterium fuses then helium would be produced
- if deuterium fuses then protons or neutrons or gamma rays would be produced.

3 1.5×10^7 °C

4 To allow time for peer review or to allow time for other scientists to check the work.

5 a They broke their agreement with Steven E. Jones.

b Two from:
- they couldn't explain how their experiment was working
- there was no control
- other scientists could not repeat the results
- there was no helium produced (which would be predicted by the theory)
- there were no neutrons produced (which would be predicted by the theory)
- there were no protons produced (which would be predicted by the theory)
- there were no gamma rays produced.

6 The thermometer was only measuring the temperature in one part of the apparatus, which got warmer during the experiment. The liquid was colder in other parts but they assumed that all of the liquid was the same temperature, meaning that when they calculated the total amount of heat energy in the whole apparatus at the end of the experiment they greatly overestimated it. *This is quite tricky. Make sure you can follow the reasoning.*

7 **WEA** *This question asks you to <u>explain</u> what a control is so make sure you explain why investigations sometimes need a control – don't just describe what a control is.*

Check your spelling, grammar and punctuation and try to include scientific words, such as 'deuterium' and 'fusion'.

You should try to include some of the points in the list below. You don't need to make all the points but you should aim to make points with the more stars. Also look at the grade booster box on page 75 and the general points about writing extended answers on page 86.

★ *you need a control to see if something is working*
★ *you could replace one thing that could then act as a control (e.g. water)*
★★ *a control is usually when the independent variable (input variable) is not applied*
★★ *a control lets you make comparisons*
★★ *use normal water (H_2O) as a control, instead of 'heavy water' (D_2O)*
★★ *or use another metal as a control, instead of palladium*
★★★ *a control allows you to form a baseline to compare with your other results to see if the changes are due only to the independent variable*
★★★ *there are two independent variables referred to in Figure B: the amount of D_2O and the volume of palladium. So replacing either of these with closely related substances will act as a suitable control (e.g. H_2O and a different metal).*

P2 GALILEO'S IDEA

1 Qualitative. *It only tells us that the objects landed together … there are no figures.*

2 100 ms. *There are 1000 milliseconds (ms) in 1 second (s). Multiply 1 s by 1000 to get the figure into ms. In this case, 0.1 s × 1000 = 100 ms. You should have also given 'ms' as the unit.*

3 Galileo's experiment has been done many times by many different people, with the same result. So his experiment is reproducible. The more reproducible a finding is, the more likely it is to be correct.

4 a $s = d/t$ so $s = 176/0.6 = 293$ cm/s or 2.93 m/s. *To get this correct you need to have written down the distance fallen by the ball at the end (176 cm) and divided this by the time taken. Each image in the photo is 0.1 seconds apart and there are six images below the starting image, so that means that the ball has taken 0.6 seconds to travel that distance. Also be careful of your units. The units that you give your answer in will depend on which units you used for the distance and the time.*

b He did not have a stopwatch/clock/watch with a small enough resolution.

5 a

Distance (cm)	Time squared (s^2)
5	0.01
20	0.04
44	0.09
78	0.16
123	0.25
176	0.36

b *A scatter graph is drawn here (see next page) because you are looking for relationships/correlations between two qualitative and continuous variables.*

You might have included error bars, which are always a nice touch but you won't lose credit for not including them.

Make sure you have:

- *a title*
- *the dependent variable on the y-axis (distance)*
- *the independent variable on the x-axis (time squared)*
- *a good scale for the y-axis so that the plotted points are well spread*
- *a y-axis scale that has even divisions*
- *a y-axis scale that is numbered*
- *a label for the y-axis with units*
- *a good scale for the x-axis so that the plotted points are well spread*
- *a label for the x-axis with units*
- *an x-axis scale that has even divisions*
- *an x-axis scale that is numbered*
- *all points accurately plotted*
- *all points neatly plotted*
- *a line of best fit drawn using a ruler in the correct area of the graph.*

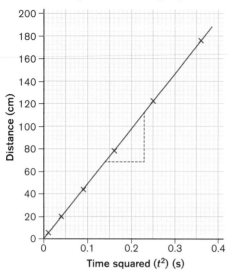

Relationship between distance and time squared for a falling object

Distance (cm) vs *Time squared (t^2) (s)*

c Distance is directly proportional to time squared, or $d \propto t^2$. *Try to get into the habit of using symbols.*

d 489 cm/s^2. The constant of proportionality is calculated from the gradient of the graph.

$$\text{gradient} = \frac{\text{change in } y}{\text{change in } x} = \frac{116 - 72}{0.23 - 0.14} = \frac{44}{0.09} = 489$$

Remember to choose points that are easy to read off your graph when you need to calculate a gradient.

6 $v^2 - u^2 = 2 \times a \times x$

v = unknown
u = starting speed = 0
a = acceleration due to gravity = 9.8 m/s^2
x = height of the tower = 56 m

Step 1: Get v as the subject.

$v^2 = (2 \times a \times x) + u^2$

So, $v = \sqrt{(2 \times a \times x) + u^2}$

Step 2: Put in the numbers.

$v = \sqrt{(2 \times 9.8 \times 56) + 0}$

$v = \sqrt{1097.6}$

v = 33 m/s (to 2 significant figures to match the number of significant figures given in the question)

7 You should try to include some of the following points, aiming for the points with the more stars.

★ flow chart drawn, with the idea coming before the testing

★ statement of Galileo's idea (gravity acts equally on all objects)
★ statement of a prediction (e.g. the objects will hit the ground together)
★ description of experiment
★★ clear statement showing that Galileo's idea about gravity acting equally on all objects is a hypothesis
★★ clear statement showing that Galileo's hypothesis became a theory when experimental results agreed with the prediction
★★ show that other predictions have been made using the theory
★★★ outline other predictions that have been made using Galileo's theory: a hammer and feather will hit the surface of the Moon at the same time when dropped from the same height, a ball and feather will hit the ground at the same time when in a vacuum

P3 PET SCANS

1 a What happens to electrons in different situations?

b If this model is correct then positively charged electrons must also exist.

2 a

Cancer number	Volume of cancer (mm³)
1	414
2	606
3	2150
4	697
5	3320
6	3050
7	1950
8	2150
9	1020
10	144

The volume of a sphere needs the equation $\frac{4}{3} \times \pi \times r^3$. Don't forget the units in this. The radii were given in mm and so the units will be mm³.

b Because a cancer is not going to be perfectly spherical.

c *You need to first state why scientists use trial runs, for example:*

- it allows you to make sure a method works or
- it allows you to work out what measurements to make in a larger investigation.

And then you need to explain your statement, for example:
- so that you don't waste time on taking measurements that do not tell you anything
- so that you don't waste money using equipment that doesn't need to be used.

d 5.50 mm – 6.25 mm. The resolution is the smallest radius that can be detected. The smallest of the cancers that could be detected by the PET scanner had a radius of 6.25 mm and the largest of those that could not be detected had a radius of 5.50 mm. The smallest one that could actually be detected must lie somewhere between these two.

3 **[WEA]** *There are some examples of answers to this question, and some comments on those answers, on pages 88–89.*

This question asks you to discuss something, which means that you need to build up an argument about whether it is a good idea or not to use PET scans. You will need to think about the benefits, drawbacks and risks and then put your ideas down in the form of an argument (see S60).

Check your spelling, grammar and punctuation and try to include scientific words, such as 'radioactive'.

You should try to include some of the points in the list below. You don't need to make all the points but you should aim to make points with the more stars. Also look at the grade booster box on page 77 and the general points about writing extended answers on page 86.

★ identify one benefit of using a PET scanner (e.g. it's good at detecting cancer, it's quick)

★ identify one drawback of using a PET scanner (e.g. expensive, not suitable for small children, not suitable for pregnant women, takes a up a lot of room)

★ identify one risk of using a PET scanner (e.g. risk of the scan causing cancer in the patient, risk of the scanner causing cancer in the operator)

★★ explain two or more benefits of using a PET scanner (e.g. can detect cancer where other techniques cannot, it's quick and so the patient can get the results quickly and treatment can be started as soon as possible)

★★ explain two or more drawbacks of using a PET scanner (e.g. expensive because the tracers cost a lot to make, not suitable for young children/pregnant women because the risks of cancer caused by the radioactive tracer are greater)

★★ explain two or more ways in which risks are reduced (e.g. amount of radioactivity used in the tracer is kept to a minimum, the operator is shielded from the equipment)

★★★ PET scans can detect small cancers that cannot be seen using other techniques, for example in Figure A the eye cancer shows up on the PET scan but not on the CT scan

★★★ only some hospitals have PET scanners, because the tracers need to be made close-by, and so you may have a travel a long way to have a PET scan

★★★ a PET scan increases your risk of getting cancer by 1 in 10 000 and the risk increases for people who are still growing, so children and pregnant women do not usually have PET scans

★★★ an argument written with the same structure as shown in S60

★★★ a clear response given to one counterargument

P4 ON THE SOCKS

1 Controlled variables: the type of socks, the footpath; uncontrolled variables: the sex of the volunteers, the ages of the volunteers, the heights of the volunteers. *There are quite a lot of these and you may be able to think up some more.*

2 a The 'no socks' group since this is the group where the independent variable is not applied.

   b Control groups are used when it is difficult to control all the variables. The results from the control group form a baseline to compare with the other results.

3 The reading of 69.4 s in the 'no socks' group is outside the general grouping of the other results.

4 a Socks group: slipperiness rating mean = 1.6, time mean = 37.7 s

   No socks group: slipperiness rating mean = 2.9, time mean = 39.6 s

   b To come up with an estimate of a measurement's true value.

   c You should have drawn one bar chart for the slipperiness and one for the mean time to descend the slope. See the next page.

You need bar charts here because the independent variable is qualitative but the dependent variable is quantitative, continuous.

You might have included error bars, which are always a nice touch but you won't lose credit for not including them.

Make sure you have:

- *a title*
- *the dependent variable on the y-axis*
- *the independent variable on the x-axis*
- *a good scale for the y-axis so that the plotted points are well spread*
- *a y-axis scale that has even divisions*
- *a y-axis scale that is numbered*
- *a label for the y-axis (with units)*
- *a label for the x-axis*
- *gaps between the x-axis bars*
- *the categories correctly labelled on the x-axis*
- *all bars accurately plotted*
- *all bars neatly drawn with a ruler (they needn't be coloured).*

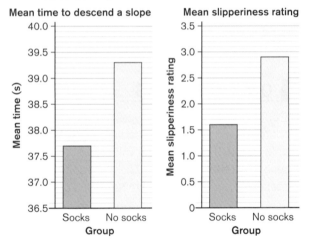

5 People might have fallen and slipped a long way, with a greater likelihood of injury. *It is generally agreed that experiments should not harm people.*

6 You could present your tally chart as grouped data (as shown below) or as a tally for each slipperiness rating (0, 0.5, 1, 1.5, 2, 2.5, etc.).

Slipperiness rating	Tally	Frequency
0 – 0.9		0
1 – 1.9	///	3
2 – 2.9	##//	5
3 – 3.9	/	1
4 – 4.9	////	4
5 – 5.9	//	2

7 So that there is no bias caused by the experimenter because the experimenter does not make the choices. *If the experimenter were allowed to choose which people used the socks and which people didn't, he/she might (for example) give the socks to people he/she thought were less likely to slip to prove that socks help. Experimenters sometimes do this unintentionally.*

8 What type of socks are best? Does someone's height determine whether socks outside shoes are helpful? Do socks actually increase the friction between ice and someone's feet? *There are quite a lot of these and you may be able to think up some more.*

9 **WEA** *This question asks you to put your ideas down in the form of an argument (see S60).*

Check your spelling, grammar and punctuation and try to include scientific words. Note that a 'hazard' and a 'risk' are different things.

You should try to include some of the points in the list below. You don't need to make all the points but you should aim to make points with the more stars. Also look at the grade booster box on page 78 and the general points about writing extended answers on page 86.

★ identify something that could cause harm (e.g. ice, flapping bit of sock)
★★ identify one or more hazards (e.g. ice, flappy piece of sock)
★★ identify one or more risks (e.g. risk of slipping on ice, tripping over flappy sock)
★★ identify one or more ways of reducing the risk (e.g. pull the sock properly over the shoe)
★★★ the data from Dr Parkin's investigation shows that people wearing socks over their shoes felt less slipperiness and got down the icy footpath quicker, and so wearing socks over shoes is a good way to reduce the risk of slipping
★★★ include a counterargument (e.g. some people might say that the study was too small to be confident of the results, some people might say that the study concentrated on people's feelings of slipperiness rather than actual quantifiable slipperiness)
★★★ with the slipperiness rating for the 'socks group' being about half that (55%) of the 'no socks group', this is good evidence that the socks helped to reduce slipperiness
★★★ a clear response given to one counterargument (e.g. some people might say

that the study concentrated on people's feelings of slipperiness rather than actual quantifiable slipperiness, but the people's perception of slipperiness was backed up by the length of time it took them to descend)

★★★ an argument written with the same structure as shown in S60.

P5 SOUND ADVICE

1 12 kHz. *Make sure you include the unit.*

2 The measurement for Lucy at 16 000 Hz because it does not follow the trend of the other points.

3 The higher the frequency, the louder the sound has to be to hear it.

4 a micropascals (thousandths of a pascal)

 b Using symbols speeds up writing things down, it makes things look clearer (and so you can understand it more quickly), and symbols can be understood all over the world no matter what language is spoken. *The best answer will make all three of these points.*

 c $2 \times 10^6 \,\mu$Pa. *Don't forget the units.*

 d $2 \times 10^6 = 2000 \times 10^3$

 $2000 \times 10^3 - 5 \times 10^3 = (2000 - 5) \times 10^3 = 1995 \times 10^3 = 1.995 \times 10^6 \,\mu$Pa

 The thing to remember is to make all the numbers have the same index. So, you could have done it the other way, and started by converting 5×10^3 (to 0.005×10^6). And don't forget the units!

5 a 17.4 kHz and 60 dB. Both Lucy and Brian will hear the 8 kHz sound at all the loudness settings. At 17.4 kHz, at 40 dB Brian can't hear it but Lucy can only just hear it so it may not be very effective. At 60 dB Brian can't hear it but Lucy can. At 100 dB Brian and Lucy can both hear it.

 b Yes. Using two people does not get enough data to draw a conclusion. Secondary data would give the shopkeeper information about the frequencies heard by hundreds/thousands of people in different age groups, without having to do the investigations.

6 This is not a valid conclusion since 25 000 Hz was not tested. *You can't really extrapolate this data because it's not a regular line (e.g. a straight line). It looks as if Brian would not be able to hear*

sounds in this range but his hearing is not necessarily the same as that of all adults and this frequency was not tested.

7 **WEA** *Notice that the question asks you to explain how you would go about doing something; it does not ask you to reach a decision. When answering questions like this, think about financial costs, effects on different groups people, effects on the environment and ethics. Then think about what you already know and what you need to find out in order to reach a decision.*

Check your spelling, grammar and punctuation and try to include scientific words, such as 'frequency'.

You should try to include some of the points in the list below. You don't need to make all the points but you should aim to make points with the more stars. Also look at the grade booster box on page 79 and the general points about writing extended answers on page 86.

★ identify a benefit of installing the HPDD (e.g. the shop may make more money if people aren't put off by youths outside it, it's more convenient than asking people to move away from the shop the whole time)

★ identify a drawback of installing the HPDD (e.g. costs money)

★ you need to know if the HPDD works

★★ in making a decision you need to balance the benefits and the drawbacks

★★ find out whether it will affect animals in the environment

★★ find out how it will affect people using the shop

★★ find out whether it is fair on people using the shop and passers-by

★★★ find out the frequencies that animals can hear (e.g. bats) and work out whether the sound might disturb them and so harm the natural environment

★★★ find out whether the sound will upset babies and cause concern for their parents (who may not be able to hear the noise)

★★★ decide whether installing it will have a good effect for more people than it has a bad effect on, and so be able to justify this as an ethical decision

1 a His hypothesis explained why planets appeared to become brighter and dimmer at different times of the year and why some planets appeared to go backwards at some times in the year. *For explanations, see the answer to question 7.*

  b The prediction that parallax would be observed for stars was shown to be correct. The hypothesis could now explain all the observations and had a lot of evidence to support it so it became a theory.

2 The resolution of the instruments was not good enough to detect the parallax in the 16th century. *This was a time before even telescopes had been invented.*

3 1/60 × 1/60 = 1/3600

4 a 3.15/1409 = 0.002236 arcseconds/pixel. Pluto's width is 48 pixels and so, 48 × 0.002236 = 0.1073 arcseconds. *The answer is given to 4 significant figures because the pixels at the start are given to 4 significant figures. If you have given your answer to 3 significant figures, this will still be correct. Don't forget the units.*

  🔺 *Although the question didn't ask you to show your working, you won't get any credit for an answer that is wrong and shows no working.*

  b 1073/10000. *If you have used fewer significant figures in your answer you may have got 107/1000.*

  c 0.1073 × 1/60 = 0.1073 × 0.0167 = 0.001791 arcminutes or *0.00179 to 3 significant figures.*

5 a You need to have drawn a scatter graph because you are looking for relationships/correlations between two qualitative and continuous variables.

  Make sure you have:
  • *a title*
  • *the dependent variable on the y-axis (distance)*
  • *the independent variable on the x-axis (parallax)*
  • *a good scale for the y-axis so that the plotted points are well spread*
  • *a y-axis scale that has even divisions*
  • *a y-axis scale that is numbered*
  • *a label for the y-axis with units*
  • *a good scale for the x-axis so that the plotted points are well spread*
  • *a label for the x-axis with units*

  • *an x-axis scale that has even divisions*
  • *an x-axis scale that is numbered*
  • *all points accurately plotted*
  • *all points neatly plotted.*

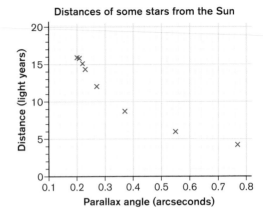

Distances of some stars from the Sun

  b They are inversely proportional. *The graph is a steady curve that slopes down to the right.*

6 a It would shorten the calculated distances between the Sun and the stars. *For inversely proportional variables, as one variable increases the other decreases and vice versa.*

  b Systematic. This is because all the parallaxes were overestimated.

7 19 °C

8 **[WEA]** *You are asked to explain things in this question. So don't just describe the observations.*

  Check your spelling, grammar and punctuation and try to include scientific words, such as 'parallax' and 'theory'.

  You should try to include some of the points in the list below. You don't need to make all the points but you should aim to make points with the more stars. Also look at the grade booster box on page 81 and the general points about writing extended answers on page 86.

  ★ identify two or more observations (e.g. the planets get brighter and dimmer, stars will show parallax, some planets appear to go backwards at certain times of the year)
  ★ state Copernicus' theory that the Earth went around the Sun
  ★★ if the Earth is further from a planet it will look dimmer, and as the Earth moves it gets closer to a planet making it look brighter
  ★★ if you look at a fixed object from different places, its position seems to change against the background. Since the Earth is moving, you expect to see closer stars moving across the background of more distance stars

★★ the reason why some planets appear to go backwards against the pattern of stars is a parallax effect. *Look at Figure B and imagine that the star is a planet, and the Earth moves down the page to viewpoint B. The 'planet' appears to move up the page against the background of stars as this happens. However, once the Earth starts to move back up towards viewpoint A, the 'planet' will now appear to be going in the other direction.*

★★★ if everything were revolving around the Earth, the Earth would not be moving and so there would be no parallax of the stars

★★★ if everything were revolving around the Earth, all the planets would be seen to be circling around the Earth and would not go backwards

★★★ if everything were revolving around the Earth, the planets would remain at a constant distance and so would not get brighter and dimmer

P7 SPF

1 a By using a sunscreen. *This is the expected answer but you'll get credit for other sensible suggestions like wearing a hat/clothes, not going outside, sitting in the shade.*

b People are more concerned about hazards that can cause long-term harm than those that cause short-term harm, or cancer is perceived to be more dangerous.

2 a The SPF number of the sunscreen.

b Two from: the amount of sunscreen; the thickness of the sunscreen smear; the brand/type of sunscreen; the amount of UVB radiation reaching each part of the card.

3 150 mins. *The SPF is a ratio, so $15 = x/10$. So, $15 \times 10 = x = 150$. However, note that this is only if you apply enough of the sunscreen (which many people don't) and if you don't go swimming or sweat too much, which makes the sunscreen come off your skin.*

4 The ones of the SPF 30 sunscreen. *These three readings are the most different from one another.*

5 a SPF 4 = 73; SPF 8 = 133; SPF 15 = 276; SPF 30 = 676; SPF 50 = 1915

b You need to have drawn a scatter graph because you are looking for relationships/correlations between two qualitative and continuous variables.

You might have included error bars, which are always a nice touch but you won't lose credit for not including them.

Make sure you have:
- *a title*
- *the dependent variable on the y-axis (time for colour change)*
- *the independent variable on the x-axis (spf number)*
- *a good scale for the y-axis so that the plotted points are well spread*
- *a y-axis scale that has even divisions*
- *a y-axis scale that is numbered*
- *a label for the y-axis with units*
- *a good scale for the x-axis so that the plotted points are well spread*
- *a label for the x-axis with units*
- *an x-axis scale that has even divisions*
- *an x-axis scale that is numbered*
- *all points accurately plotted*
- *all points neatly plotted*
- *(a line of best fit drawn using a ruler, if you've done one).*

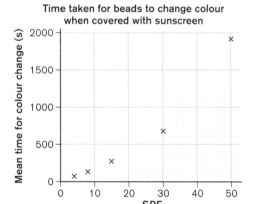

Time taken for beads to change colour when covered with sunscreen

c The greater the SPF number, the longer the bead takes to change colour.

6 You should make the following main points.
- The investigation is valid since it is measuring the effect of blocking UVB rays getting to the skin, which would otherwise cause skin damage.
- The measurements are valid and measure how well the sunscreen blocks the UVB rays.
- However, some of the results are not very precise and it would be better to have done more repeat measurements, particularly for the SPF 30 sunscreen.
- The results are accurate enough to draw a conclusion because all of the results, even the less precise set, show a gradual increase in UVB protection as the SPF number increases.

- Apart from one set of the results the measurements were repeatable.
- The results are unbiased but no conclusion has been drawn.

7 **WEA** *This question asks you to give your own opinion and to explain how you came to this point of view from the information given in both P7 and B9.*

Check your spelling, grammar and punctuation and try to include scientific words, such as 'ultraviolet', 'rays', 'ozone', 'SPF', 'hypothesis'. You should try to include some of the points in the list below. You don't need to make all the points but you should aim to make points with the more stars. Also look at the grade booster box on page 82 and the general points about writing extended answers on page 86.

★ state a reason for sunbathing (e.g. lets the skin make vitamin D, relieves stress)

★ state a reason against sunbathing (e.g. causes sunburn, too much sun can damage the skin permanently, can cause skin cancer)

★ use scientific terms (e.g. ultraviolet rays, ozone, SPF, hypothesis)

★★ use information from both pages, including references to the data and graphs

★★ answer structured in paragraphs

★★ scientific terms used and spelled correctly

★★★ information synthesised (put together) from both pages (e.g. use the risks on page 62 (B9) together with ways of reducing the risk from page 82 (P7); describe the effects of UVA rays from page 82 (P7) along with the effects of UVB rays from page 62 (B9))

★★★ answer structured in the form of a statement of opinion, which is justified by paragraphs each making a central point that is backed up with evidence from the pages, and finishes with a clear concluding paragraph

P8 SPEED LIMITS

1 a km/h

  b Compound units are formed from two or more other units.

2 a Total travel time is (10 mins 4 s) − (3 mins 2 s) = 604 s − 182 s = 422 s. *You can ignore the 14 hours in this case. The 14 refers to 2 pm.*

  b 422 s = 422 ÷ 60 mins = 7.03 mins = 7.03 ÷ 60 hours = 0.117 hours. Speed is distance ÷ time, so 7 ÷ 0.117 = 59.8 mph. *Don't forget the units.*

c 10% of 50 = 10/100 × 50 = 5. Cars going above 50 + 5 + 2 = 57 mph will get a fine. Yes, this driver will be fined.

3 15 mins = 0.25 hours. 40 × 0.25 = 10 km. *Don't forget the units! A common mistake here is to forget to convert the minutes into hours. You need to use the same units of time for both the time and the speed. And here, you also need to re-arrange the equation so that it becomes distance = speed × time.*

4 0.279 = 27.9%

5 a This is the value at which 85% of the data is cut off. In this case it is the speed at or below which 85% of drivers are travelling.

  b At the 50th percentile (which divides the data in two).

6 You need to draw a bar chart of grouped continuous data here because the independent (input variable) is continuous but has been grouped.

Make sure you have:

- *a title*
- *the dependent variable on the y-axis*
- *the independent variable on the x-axis*
- *a good scale for the y-axis so that the plotted points are well spread*
- *a y-axis scale that has even divisions*
- *a y-axis scale that is numbered*
- *a label for the y-axis*
- *a good scale for the x-axis so that the data is well spread*
- *a label for the x-axis*
- *no gaps between the x-axis bars*
- *an x-axis scale that has even divisions*
- *the categories correctly labelled on the x-axis (you may have labelled the groups as shown or drawn a scale)*
- *all bars accurately plotted*
- *all bars neatly drawn with a ruler.*

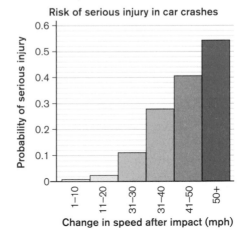

Risk of serious injury in car crashes

Probability of serious injury (y-axis: 0, 0.1, 0.2, 0.3, 0.4, 0.5, 0.6)

Change in speed after impact (mph) (x-axis: 1–10, 11–20, 31–30, 31–40, 41–50, 50+)

7 **[WEA]** *This question asks you to discuss something, which means that you need to identify the key points (points for and against raising the speed limit) and then build up an argument (see S60) to say whether you think this is a good idea or not. You will need to think about the benefits, drawbacks and risks.*

There is no right or wrong answer for this question. It is how your use the information and structure your answer that matters. Check your spelling, grammar and punctuation and try to include scientific words.

You should try to include some of the points in the list below. You don't need to make all the points but you should aim to make points with the more stars. Also look at the grade booster box on page 83 and the general points about writing extended answers on page 86.

★ a benefit is that people get to places more quickly

★ a drawback is that if cars go too fast it might scare new drivers

★ there may be a greater risk of crashing if people are encouraged to go too fast

★★ there is a greater risk of more severe injury, the faster a car is going when it crashes

★★ many drivers already go faster than 70 mph, so there may not be too much of a change if the limit is increased

★★★ the graph shows that the risk of crashing is reduced if people are driving at the 85th percentile speed (which the text tells you is 80 mph)

★★★ the reduction in the risk of crashing between the mean motorway speed and the 85th percentile (80 mph) is not great

★★★ increasing the speed limit may shift the distribution of speeds towards the 85th percentile speed (80 mph), meaning that more drivers are in the area at the end of the graph and so have a much increased risk of crashing

P9 MEASURING DISTANCES

1 a He measured the sensitivity of the LDR and the photodiode.

b Two from:
- to work out the range of measurements needed
- to work out the interval needed between the measurements
- to work out how many measurements to take
- to work out how many repeat readings to take
- to make sure that you don't make unnecessary readings in the full investigation
- to make sure a method works, and you have the correct apparatus
- to make sure that you have accurate enough measuring apparatus
- to make sure that your investigation is safe.

2 $0.2 - 1.0 = 0.8$ m. *Remember that the independent (input) variable is the variable that the experimenter changes.*

3 *Your circuit diagram doesn't need to look exactly like this but it must have the three symbols shown (in any order) and must be drawn using a ruler!*

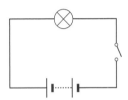

4 a $3.4\,V \pm 0.1\,V$

b $V = I \times R$
so $R = V/I$
$= 3.4/0.1$
$= 34\,\Omega$

c $P = I \times V$
$P = 0.1 \times 3.4$
$= 0.34\,W$
(or you could use $P = I^2 \times R$)

5 The sensitivity is how small a change something can detect. The LDR could detect smaller changes in light level than the photodiode.

6 $2\,V$

7 a $(0.22 - 0.21) \div 0.21 = 0.048$ which is $0.048 \times 100 = 4.8\%$
$(0.46 - 0.43) \div 0.43 = 0.07$ which is $0.07 \times 100 = 7\%$
$(0.75 - 0.79) \div 0.79 = -0.051$ which is $-0.051 \times 100 = -5.1\%$ *This is a negative number because the prediction is lower than the actual value. However, if you have calculated it as a positive number that's also fine. Using the negatives just helps you to answer the next question.*

b Random error. *It's random because with a systematic error all the predictions would be* higher or lower than the actual values (not a mixture of higher and lower).

c Not reading the voltages/distances correctly. Or light sometimes getting to the LDR from other sources (not just by reflection from the board). This investigation would need to be done in a dark room but it's difficult to make the room totally dark the whole time.

8 [WEA] *This question asks you to evaluate the investigation. So you need to say how good or poor the investigation is based on the list of things to check given in S54.*

Check your spelling, grammar and punctuation and try to include scientific words, such as the correct terms for the units.

You should try to include some of the points in the list below. You don't need to make all the points but you should aim to make points with the more stars. Also look at the grade booster box on page 84 and the general points about writing extended answers on page 86.

★ it would have been better if the measurements had been repeated

★ the conclusion matches the hypothesis

★ the measuring devices were accurate enough for this investigation

★★ the investigation is valid: the results allow you to answer the question

★★ the conclusion is valid because it is only drawn from the results

★★ the data for the graph was of good quality because, although there were no repeated measurements, the results obtained for the graph in Figure B all fit neatly on a regular curve

★★★ the conclusion would be better (more certain) if more measurements had been taken for the second part of the investigation. A set of only three measurements does not provide very strong evidence that this way of measuring distances is not accurate

★★★ it would have been better to draw a curve of best fit through the points on the graph because the variables seem to have an inverse relationship

P10 SPRINGS

1 a 19.3 − 9.8 = 9.5 cm = 0.095 m

b 95 mm

2 The 50 g set, because they are where the repeated readings are closest together.

3 a One of: human errors (random) in reading the ruler scale; inaccuracy in measurement of the masses used; the spring might have moved slightly against the ruler; the spring might have touched the ruler (preventing it from stretching fully).

b If the ruler slipped / was not in the same position for repeated readings.

4 a

Mass added (g)	Mean length of spring (cm)
0	9.8
50	19.3
100	29.3 (3 s.f.)
150	39.1 (3 s.f.)
200	49.1 (3 s.f.)
250	58.7 (3 s.f.)
300	68.4
350	78.3
400	87.7
450	97.2

b 97.2 ± 0.1 cm

c scatter graph with a line of best fit

d 9.8 cm

e Gradient ≈ 0.2

f $y = 0.2x + 9.8$

5 a When mass is applied the extension increases by the same percentage as the mass has increased.

b ∝

c 0.05 × 9.8 = 0.49 N. (Remember that the mass has to be in kilograms for this equation.)

6 a Count up the squares or use the area of a triangle (0.5 × base × height).

b Working shown to convert all the masses into weights. Working shown to calculate extensions. Scatter graph drawn with a line of best fit. Clear working shown to calculate the area under the graph.

Glossary

Identifiers	Word	Definition
S50, S54	accuracy	How close a measurement (or set of measurements) is to the real value of something.
S19, S20	algebraic equation	An equation that uses letters to represent quantities.
S6, S15, S39, S48, S54	anomalous result	A measurement that does not fit the same pattern as other measurements from the same experiment.
S32, S57	area	The amount of surface that a shape has.
S60	argument	Telling people what you believe and why, usually with reasons why you do not agree with others.
S58	assess	Command word in questions that means the same as 'evaluate'.
S54	assumption	Something that everyone accepts as true so you don't test it when doing an investigation, e.g. the Earth is like a ball.
S8, S9, S23	bar chart	Chart showing solid columns, used to present data. It is often used when the independent variable is qualitative.
S36	benefit	Something good that could come from an action or device.
S12, S39, S50, S54	bias	If evidence is shifted in a particular direction it shows bias.
S58	calculate	Command word in questions, meaning work out an answer using numbers. Always show your working. Always put in the units.
S9, S41	categoric data	Data that is not in the form of numbers. Also called 'qualitative data'.
S16	causal correlation	A correlation caused by the independent variable directly affecting the dependent variable.
S32	circumference	The distance around the outside of a circle.
S58	compare	Command word in questions, meaning describe the differences/similarities between things or their advantages/drawbacks.
S58	complete	Command word in questions, meaning fill in answers in a space or finish writing a sentence.
S21, S56	compound measure	Another term for 'compound unit'.
S21, S56	compound unit	A unit of measurement made up of more than one unit (e.g. m/s).
S39, S52, S54, S60	conclusion	A decision made after looking at all the evidence.
S25, S27	constant of proportionality	Symbol is often 'm'. The amount that the values on the x-axis are multiplied by to become the same as the values on the y-axis so that $y = mx$.

Identifiers	Word	Definition
S10, S15, S41	continuous data	Data in which each value can be any number between two limits.
S46	control	In an experiment, a control uses exactly the same set-up as the main part of the experiment but without the independent variable.
S46	control experiment	An experiment in which a control (or control group) is used.
S46	control group	A control for an experiment that consists of a group of things (usually organisms).
S7, S43, S53	control variable	A variable that needs to be controlled in an experiment, otherwise it will affect the results.
S16	correlation	A link between two variables, so that when one changes so does the other one.
S60	counterargument	A reason for not agreeing with an argument.
S15	curve of best fit	A regular curve drawn through a set of points on a graph so that about half the points are above the line and half are below it.
S41	data	Numbers, words, etc. that can be organised to give information.
S14	decile	When data is divided into 10 equal parts: the upper limit of the first 10th of the data is the first decile; the upper limit of the second 10th is the second decile, etc.
S1, S3	decimal number	A number shown as a single line of digits with a decimal point separating whole numbers from tenths, hundredths and so on.
S58	define	Command word in questions, meaning state briefly what something means
S3	denominator	The number below (or after) the line in a fraction.
S7, S9, S16, S23, S24, S26, S27, S43	dependent variable	A variable that depends on the changes of another variable. This is the variable that you measure in an experiment.
S58	describe	Command word in questions, meaning recall facts in an accurate way or say what a diagram or graph shows (e.g. what trend you can see).
S32	diameter	The distance going through the centre of circle from one side to the other.
S25, S27	directly proportional	Two variables are directly proportional when one increases and the other then increases by the same percentage. This written as $A \propto B$.
S9, S41	discrete data	Data in which each value can only be one of a limited choice of numbers.
S58	discuss	Command word in questions, which means build up an argument about an issue.

Identifiers	Word	Definition
S36, S37	drawback	Something undesirable that could come from an action or device.
S9, S15	error bar	A line drawn up and down from an averaged point on a graph to show the range of readings for that point.
S4, S12, S58	estimate	Command word in questions, meaning make a rough calculation.
S35, S38	ethics	Actions or ideas that a group of people agree are right or wrong. A country's laws are based on ethics.
S39, S54, S58	evaluate	Assess how well something does or has done its job. Used as a command word in questions, when you have to say how good or poor something is based on a series of points (criteria). If you are asked to evaluate more than one thing then you need to compare the things (see 'compare' above) and then state which of the things is best, with reasons why you think that.
S39, S54	evaluation	An assessment of how well something does or has done its job.
S34	evidence	Information used to support an idea or show that it is wrong.
S58	explain	Command word in questions, meaning state the reasons why something happens.
S15, S25	extrapolation	Using a part of some data to estimate values outside this part of the data.
S43, S53	fair test	An experiment in which all the control variables are successfully controlled so that the only factor that affects the dependent variable is the independent variable.
S23, S33, S59	flow chart	Set of boxes with arrows showing how to move through the steps of a process, including any choices that may need to be made.
S3	fraction	A number shown as two parts with a line between them. The digit below (or after) the line is the total number of possible parts. The digit above (or in front of) the line is the actual number of parts.
S8, S10	frequency	The number of occurrences of something in a certain time or in a certain area.
S8, S10	frequency diagram	Any chart or graph that shows frequency of data.
S8	frequency table	A table used to keep a count (a tally) of frequencies.
S58	give	Command word in questions, meaning state a fact or example.
S25, S27	gradient	The value that describes the slope of a straight line. Calculated by picking points, and calculating the difference in the values on the y-axis between these two points, and then difference between the two x-values. The gradient is the difference in y divided by the difference in x.

Identifiers	Word	Definition
S37, S47	hazard	Something that can cause harm.
S8, S10, S23	histogram	A frequency diagram showing frequency as the area of a bar in a bar chart that groups together data on a continuous scale. The y-axis shows frequency density.
S51	human error	Errors in measurements caused by the people making the measurements.
S31, S34, S40, S52, S53	hypothesis	A scientific idea that can be tested.
S58	identify	Command word in questions, meaning look at some data or text and pick out a certain part.
S58	illustrate	Command word in questions, meaning give examples in an explanation or description.
S7, S9, S16, S23, S24, S26, S27, S43	independent variable	A variable that does not depend on changes in other variables. This is the variable that you change in an experiment.
S57	index	A small number written above another number or a unit to show how many times it should be multiplied by itself. Plural = indices. Also called the 'power'.
S57	index form	When an index is shown next to a unit or number, it is said to be in index form (e.g. m^2).
S7, S9, S16, S23, S24, S26, S27, S43	input variable	Another term for 'independent variable'.
S1, S3	integer	A number that does not contain fractions, nor a decimal point.
S27	intercept	The point at which a line on a graph crosses the y-axis.
S25	inversely proportional	Two variables are inversely proportional when one changes and the other changes in the opposite way, so that $A \propto 1/B$.
S39	journal	Scientific magazine in which papers are published.
S58	justify	Command word in questions, meaning evaluate (see 'evaluate' above) something by providing evidence for your choice of which is best.
S34	kinetic theory	Theory based on the idea that all matter is composed of particles that move.
S23, S24, S25	line graph	Graph used to present data in which both the independent and dependent variables are in the form of continuous data. Points are joined together with straight lines.
S15	line of best fit	A straight line drawn through a set of points on a scatter graph so that about half the points are above the line and half are below it.
S58	list	Command word in questions, meaning write down key points in a brief way.

Identifiers	Word	Definition
S32	mathematical constant	A number that does not change.
S6, S14	mean	The best value of a measurement, found by repeating the measurement, adding the results together and dividing the total by the number of repeats.
S14	median	The middle value of an ordered data set.
S14	mode	The most frequent item in a set of data.
S31	model	Representing a thing or a process in a way that makes it easier for us to understand. Models usually simplify the real nature of something.
S35, S38	morals	Things that you personally believe are right or wrong.
S58	name	Command word in questions, meaning state a fact or give an example.
S3	numerator	The number above (or in front of) the line in a fraction.
S6, S15, S48	outlier	Another term for 'anomalous result'.
S58	outline	Command word in questions, meaning state the main points of an argument or of how something happens.
S7, S9, S16, S23, S24, S26, S27, S43	output variable	Another term for 'dependent variable'.
S39	paper	Scientific report written by scientists to tell others about their research. Papers are published in journals.
S39	peer review	Process in which papers are checked by scientists.
S3	percentage	A fraction in which the denominator is 100. Uses the symbol %.
S3	percentage change	The difference between an initial value and a later value expressed as a percentage of the initial value.
S14	percentile	A value at which a certain percentage of some data is cut off. So the 24th percentile cuts off the first 24% of the values.
S32	perimeter	The distance around the outside edge of a shape.
S32	pi	The ratio of any circle's circumference to its diameter.
S11, S23	pie chart	Diagram in which the different proportions of something are shown as slices of a circle.
S57	power	Another term for 'index'.
S49, S50, S54	precision	How closely grouped together a set of measurements are.
S31, S40	prediction	Saying what you think the results of an experiment will be.
S42	primary data	Data that you collect, e.g. by doing an investigation.
S42	primary evidence	The same as 'primary data'.
S13	probability	The chance of something happening, shown as a fraction, a decimal or a percentage.

Identifiers	Word	Definition
S38	public enquiry	Series of meetings that anyone can attend and contribute to, which decides for or against a plan.
S9, S41	qualitative data	Data that is not in the form of numbers. Also called 'categoric data'.
S9, S41	quantitative data	Data that is in the form of numbers.
S14	quartile	When data is divided into four equal parts: the first quartile of the data is the upper limit of the first quarter part; the second quartile is the upper limit of the second quarter, etc.
S32	radius	The distance from the middle of a circle to its edge.
S12, S13	random	Something that is done without any regularity or conscious thought and is impossible to predict.
S51	random error	Errors in measurements that have no pattern to them.
S6, S45	range	In maths it is the calculated difference between the highest and lowest measured values in an experiment (usually ignoring anomalous results). In science, we often give a range in terms of the highest and lowest values.
S3, S32	ratio	A comparison between two numbers, usually shown in the form x:y.
S38, S39, S49, S54	repeatable	Results that have similar values when repeated by the same experimenter.
S38, S39, S49, S54	reproducible	Results that have similar values when repeated by the different experimenters.
S45, S51	resolution	The smallest change that a measuring device can detect.
S36, S37, S47	risk	The chance of harm occurring from a certain hazard or drawback.
S4	round	Make a number simpler but keeping its value close to what it was.
S12, S37	sample	A small part of a collection of data. Scientists might never collect all the data but just collect a sample, which they then use to estimate what the rest of the data is like.
S15, S16, S23	scatter diagram	Graph used to find correlations between two quantitative variables. Lines or curves of best fit are often drawn through the points.
S15, S16, S23	scatter graph	See 'scatter diagram'.
S33	scientific method	A series of steps that scientists take to show whether a scientific idea is right or wrong.
S42	secondary data	Data that you use but that has been collected by other people.
S42	secondary evidence	The same as 'secondary data'.
S51	sensitivity	Another term for 'resolution'.
S21, S55	SI system	System of units used by most scientists around the world.

Identifiers	Word	Definition
S5, S6	significant figures	The number of digits in a value that actually show the size of that value (all the other digits being zeros).
S27	slope	Another word for 'gradient'.
S2	standard form	A way of writing very small or very large numbers using powers of 10 (the number 10 with an index).
S58	state	Command word in questions, meaning write a fact or an example.
S19	subject	The term in an equation that appears on its own (on one side of the equals sign).
S20	substitute	Put numbers into an algebraic equation in order to do a calculation.
S58	suggest	Command word in questions, meaning use your scientific knowledge to work out what is happening in an unfamiliar situation.
S58	summarise	Command word in questions, meaning state the main points of an argument or of how something happens.
S51	systematic error	Errors in measurements that are all shifted in a certain way.
S7, S23	table	Way of recording data in order in columns and rows.
S8	tally chart	The same as 'frequency table'.
S28	tangent	A straight line drawn on a graph that skims one point on a curve (from which a gradient can then be calculated).
S33, S34	theory	A hypothesis (or set of hypotheses) that is supported by good evidence.
S45	trial run	A cut-down version of an investigation used to work out what measurements to make in the actual investigation.
S50	uncertainty	The degree to which a measurement may vary from a stated value. Often written using a ±.
S58	use the information	Command phrase in questions, meaning use the data given to answer the question.
S52, S53, S54	valid	When something does what it is meant to do.
S41	value	A number together with something that tells you what the number means (e.g. a unit of measurement).
S7, S26, S43	variable	Something that can change and have different values.
S32, S57	volume	The amount of space that a 3D shape takes up.
S1, S3	whole number	A positive number that does not contain fractions, nor a decimal point.
S58	write down	Command phrase in questions, meaning state a fact or give an example in writing.

Periodic table

Key (example):

atomic number	relative atomic mass (atomic weight)
1	1
H	
symbol	Hydrogen (name)

Main table

1	2	3	4	5	6	7	8	9	10	11	12	13	14	15	16	17	18
1 **H** Hydrogen (1)																	4 **He** Helium (2)
7 **Li** Lithium (3)	9 **Be** Beryllium (4)											11 **B** Boron (5)	12 **C** Carbon (6)	14 **N** Nitrogen (7)	16 **O** Oxygen (8)	19 **F** Fluorine (9)	20 **Ne** Neon (10)
23 **Na** Sodium (11)	24 **Mg** Magnesium (12)											27 **Al** Aluminium (13)	28 **Si** Silicon (14)	31 **P** Phosphorus (15)	32 **S** Sulphur (16)	35.5 **Cl** Chlorine (17)	40 **Ar** Argon (18)
39 **K** Potassium (19)	40 **Ca** Calcium (20)	45 **Sc** Scandium (21)	48 **Ti** Titanium (22)	51 **V** Vanadium (23)	52 **Cr** Chromium (24)	55 **Mn** Manganese (25)	56 **Fe** Iron (26)	59 **Co** Cobalt (27)	59 **Ni** Nickel (28)	64 **Cu** Copper (29)	65 **Zn** Zinc (30)	70 **Ga** Gallium (31)	73 **Ge** Germanium (32)	75 **As** Arsenic (33)	79 **Se** Selenium (34)	80 **Br** Bromine (35)	84 **Kr** Krypton (36)
85.5 **Rb** Rubidium (37)	88 **Sr** Strontium (38)	89 **Y** Yttrium (39)	91 **Zr** Zirconium (40)	93 **Nb** Niobium (41)	96 **Mo** Molybdenum (42)	99 **Tc** Technetium (43)	101 **Ru** Ruthenium (44)	103 **Rh** Rhodium (45)	106 **Pd** Palladium (46)	108 **Ag** Silver (47)	112 **Cd** Cadmium (48)	115 **In** Indium (49)	119 **Sn** Tin (50)	122 **Sb** Antimony (51)	128 **Te** Tellurium (52)	127 **I** Iodine (53)	131 **Xe** Xenon (54)
133 **Cs** Caesium (55)	137 **Ba** Barium (56)	139 **La** Lanthanum (57)	178.5 **Hf** Hafnium (72)	181 **Ta** Tantalum (73)	184 **W** Tungsten (74)	186 **Re** Rhenium (75)	190 **Os** Osmium (76)	192 **Ir** Iridium (77)	195 **Pt** Platinum (78)	197 **Au** Gold (79)	201 **Hg** Mercury (80)	204 **Tl** Thallium (81)	207 **Pb** Lead (82)	209 **Bi** Bismuth (83)	210 **Po** Polonium (84)	210 **At** Astatine (85)	222 **Rn** Radon (86)
223 **Fr** Francium (87)	226 **Ra** Radium (88)	227 **Ac** Actinium (89)	261 **Rf** Rutherfordium (104)	262 **Db** Dubnium (105)	263 **Sg** Seaborgium (106)	262 **Bh** Bohrium (107)	265 **Hs** Hassium (108)	266 **Mt** Meitnerium (109)	269 **Ds** Darmstadtium (110)	272 **Rg** Roentgenium (111)	285 **Cn** Copernicium (112)	286 **Uut** Ununtrium (113)	289 **Fl** Flerovium (114)	289 **Uup** Ununpentium (115)	293 **Lv** Livermorium (116)	294 **Uus** Ununseptium (117)	294 **Uuo** Ununoctium (118)

Lanthanide series

140 **Ce** Cerium (58)	141 **Pr** Praseodymium (59)	144 **Nd** Neodymium (60)	145 **Pm** Promethium (61)	150 **Sm** Samarium (62)	152 **Eu** Europium (63)	157 **Gd** Gadolinium (64)	159 **Tb** Terbium (65)	162 **Dy** Dysprosium (66)	165 **Ho** Holmium (67)	167 **Er** Erbium (68)	169 **Tm** Thulium (69)	173 **Yb** Ytterbium (70)	175 **Lu** Lutetium (71)

Actinide series

232 **Th** Thorium (90)	231 **Pa** Protactinium (91)	238 **U** Uranium (92)	237 **Np** Neptunium (93)	244 **Pu** Plutonium (94)	243 **Am** Americium (95)	247 **Cm** Curium (96)	247 **Bk** Berkelium (97)	251 **Cf** Californium (98)	252 **Es** Einsteinium (99)	257 **Fm** Fermium (100)	258 **Md** Mendelevium (101)	259 **No** Nobelium (102)	262 **Lr** Lawrencium (103)

The names and symbols for the elements are agreed by the International Union of Pure and Applied Chemistry (IUPAC).

Physics equations

You need to be able to recall and apply the following equations, using standard SI units. Some of these may be for Higher Tier papers only or for GCSE Physics only, depending on which exam board you are following.

distance travelled = speed × time	$s = vt$
acceleration = $\dfrac{\text{change in velocity}}{\text{time}}$	$a = \dfrac{v - u}{t}$
force = mass × acceleration	$F = ma$
weight = mass × gravitational field strength (g)	$W = mg$
momentum = mass × velocity	$p = mv$
force exerted on a spring = spring constant × extension	$F = ke$
moment of a force = force × distance (normal to the direction of the force)	
kinetic energy = 0.5 × mass × (speed)^2	$E_k = \frac{1}{2}mv^2$
gravitational potential energy = mass × gravitational field strength (g) × height	$E_p = mgh$
work done = force × distance (along the line of action of the force)	$W = Fs$
power = $\dfrac{\text{work done}}{\text{time}}$ power = $\dfrac{\text{energy transferred}}{\text{time}}$	$P = \dfrac{W}{t}$ $P = \dfrac{E}{t}$
efficiency = $\dfrac{\text{output energy transfer}}{\text{input energy transfer}}$ efficiency = $\dfrac{\text{useful power output}}{\text{total power input}}$	
wave speed = frequency × wavelength	$v = f\lambda$
charge flow = current × time	$Q = It$
potential difference = current × resistance	$V = IR$
electrical power = potential difference × current	$P = VI$
electrical power = (current)^2 × resistance	$P = I^2R$
energy transferred = power × time	$E = Pt$
energy transferred = charge flow × potential difference	$E = QV$
energy transferred = current × potential difference × time	$E = IVt$
density = $\dfrac{\text{mass}}{\text{volume}}$	$\rho = \dfrac{m}{V}$
pressure = $\dfrac{\text{force normal to surface}}{\text{area of surface}}$	$P = \dfrac{F}{A}$

You do not need to be able to recall the following equations but you will need to select and apply them. Some of these may be for Higher Tier papers only or for GCSE Physics only, depending on which exam board you are following.

(final velocity)^2 – (initial velocity)^2 = 2 × acceleration × distance	$v^2 - u^2 = 2as$
force = $\dfrac{\text{change in momentum}}{\text{time taken}}$	$F = \dfrac{m\,\Delta v}{\Delta t}$
pressure due to a column of liquid = height of column × density of liquid × gravitational field strength (*g*)	$P = h\rho g$
change in thermal energy = mass × specific heat capacity × change in temperature	$\Delta E = mc\,\Delta\theta$
thermal energy for a change of state = mass × specific latent heat	$E = mL$
energy transferred in stretching (elastic potential energy) = 0.5 × spring constant × (extension)^2	$E_e = \frac{1}{2}ke^2$
force on a conductor (at right angles to a magnetic field) carrying a current = magnetic flux density × current × length	$F = BIl$
potential difference across primary coil × current in primary coil = potential difference across secondary coil × current in secondary coil	$V_p \times I_p = V_s \times I_s$
$\dfrac{\text{potential difference across primary coil}}{\text{potential difference across secondary coil}}$ $= \dfrac{\text{number of turns in primary coil}}{\text{number of turns in secondary coil}}$	$\dfrac{V_p}{V_s} = \dfrac{N_p}{N_s}$
for gases: pressure × volume = constant (for a given mass of gas and at a constant temperature)	$PV = constant$
period = $\dfrac{1}{\text{frequency}}$	$T = \dfrac{1}{f}$
magnification = $\dfrac{\text{image height}}{\text{object height}}$	